SpringerBriefs in Water Science and Technology

SpringerBriefs in Water Science and Technology present concise summaries of cutting-edge research and practical applications. The series focuses on interdisciplinary research bridging between science, engineering applications and management aspects of water. Featuring compact volumes of 50 to 125 pages (approx. 20,000–70,000 words), the series covers a wide range of content from professional to academic such as:

- Literature reviews
- In-depth case studies
- Bridges between new research results
- Snapshots of hot and/or emerging topics

Topics covered are for example the movement, distribution and quality of freshwater; water resources; the quality and pollution of water and its influence on health; and the water industry including drinking water, wastewater, and desalination services and technologies.

Both solicited and unsolicited manuscripts are considered for publication in this series.

Joan Nyika · Megersa Olumana Dinka

The Silent Wastewater Problem in Sub-Saharan Africa

Joan Nyika
Department of Geosciences
and the Environment
School of Physics and Earth Sciences
Technical University of Kenya
Nairobi, Kenya

Megersa Olumana Dinka
University of Johannesburg
Auckland Park, South Africa

ISSN 2194-7244　　　　　　ISSN 2194-7252　(electronic)
SpringerBriefs in Water Science and Technology
ISBN 978-3-031-90142-3　　　ISBN 978-3-031-90143-0　(eBook)
https://doi.org/10.1007/978-3-031-90143-0

© The Editor(s) (if applicable) and The Author(s), under exclusive license to Springer Nature Switzerland AG 2025

This work is subject to copyright. All rights are solely and exclusively licensed by the Publisher, whether the whole or part of the material is concerned, specifically the rights of translation, reprinting, reuse of illustrations, recitation, broadcasting, reproduction on microfilms or in any other physical way, and transmission or information storage and retrieval, electronic adaptation, computer software, or by similar or dissimilar methodology now known or hereafter developed.
The use of general descriptive names, registered names, trademarks, service marks, etc. in this publication does not imply, even in the absence of a specific statement, that such names are exempt from the relevant protective laws and regulations and therefore free for general use.
The publisher, the authors and the editors are safe to assume that the advice and information in this book are believed to be true and accurate at the date of publication. Neither the publisher nor the authors or the editors give a warranty, expressed or implied, with respect to the material contained herein or for any errors or omissions that may have been made. The publisher remains neutral with regard to jurisdictional claims in published maps and institutional affiliations.

This Springer imprint is published by the registered company Springer Nature Switzerland AG
The registered company address is: Gewerbestrasse 11, 6330 Cham, Switzerland

If disposing of this product, please recycle the paper.

Competing Interests The authors have no competing interests to declare that are relevant to the content of this manuscript.

Contents

1 **The Global Challenge of Wastewater and Its Mismanagement** 1
   1.1 Introduction 1
   1.2 Sources of Wastewater 3
   1.3 Global Trends on Wastewater Production and Management 5
   1.4 Wastewater Characteristics and Composition 9
      1.4.1 Wastewater Characteristics 9
      1.4.2 Composition of Wastewater 9
   1.5 Management of Wastewater 13
   1.6 Effects of Poor Wastewater Management 15
      1.6.1 Health Impacts 15
      1.6.2 Environmental Impacts 18
      1.6.3 Socio-economic Impacts 20
   1.7 Conclusion 21
   References 21

2 **Wastewater Generation and Management Patterns in Sub-Saharan Africa** 27
   2.1 Introduction 27
   2.2 Overview of Sub-Saharan Africa Region 29
   2.3 Wastewater in Sub-Saharan Africa Region 30
   2.4 Case Studies of Wastewater Management in SSA Region 36
      2.4.1 Benin 37
      2.4.2 Ethiopia 38
      2.4.3 Malawi 39
      2.4.4 Kenya 40
      2.4.5 South Africa 40
   2.5 Discussion and Conclusion 41
   References 42

3 Wastewater Treatment and Management in SSA ... 47
- 3.1 Introduction ... 47
- 3.2 Wastewater Infrastructure in SSA ... 49
- 3.3 Wastewater Treatment Techniques ... 51
  - 3.3.1 Septic Systems ... 52
  - 3.3.2 Waste Stabilization Ponds ... 54
  - 3.3.3 Conventional Activated Sludge ... 56
  - 3.3.4 Trickling Filters ... 57
  - 3.3.5 Constructed Wetlands ... 57
  - 3.3.6 Composting Toilets ... 58
  - 3.3.7 Biodigesters ... 58
  - 3.3.8 Membrane Systems ... 59
- 3.4 Wastewater Treatment Capacity ... 60
- 3.5 Conclusion ... 61
- References ... 61

4 Challenges Facing Wastewater Management in SSA ... 67
- 4.1 Introduction ... 67
  - 4.1.1 Challenges in Wastewater Management in SSA ... 68
  - 4.1.2 Insufficient WW Infrastructure ... 69
  - 4.1.3 Financial Unsustainability ... 70
  - 4.1.4 Limited Technical and Human Capacity ... 71
  - 4.1.5 Regulatory and Governances Limitations ... 71
  - 4.1.6 Public Awareness and Education ... 73
  - 4.1.7 Poor Operations and Maintenance ... 74
  - 4.1.8 Emerging Contaminants ... 75
- 4.2 Conclusion ... 76
- References ... 77

5 Sustainable Management of Wastewater in Sub-Saharan Africa Region ... 81
- 5.1 Introduction ... 81
- 5.2 Measures to Manage Wastewater Sustainably ... 82
  - 5.2.1 Improvement of Sanitation Facilities ... 82
  - 5.2.2 Innovative and Cost Effective Wastewater Treatment Technologies ... 83
  - 5.2.3 Regulations and Standards ... 84
  - 5.2.4 Enhancement of Human Capacity ... 85
  - 5.2.5 Governance Improvement in the Wastewater Sector ... 86
  - 5.2.6 Improve Wastewater Monitoring, Data Collection and Sharing ... 87
  - 5.2.7 Enhanced Resource Reuse and Recovery from Wastewater ... 88
  - 5.2.8 Reliable Energy Supply ... 89
- 5.3 Conclusion ... 90
- References ... 91

Chapter 1
The Global Challenge of Wastewater and Its Mismanagement

Abstract This chapter explored the global trends on wastewater (WW) production and management, its sources, composition and effects on human health, the environment and socio-economy. Global production rates of WW were found to be on a rising trend with generation rates expected to increase by 51% between 2015 and 2050 from 380 to 574 billion m^3 annually. The sources of the resource are diverse but mainly driven by anthropogenic activities such as industrialization, agriculture and urbanization. Composition of WW was shown to have biological, physical and chemical constituents, some of which have a non-biodegradable and bioaccumulative nature once introduced in food chains. The effects of discharge and reuse of the resource in its raw form results to the introduction of toxicities and diseases in humans, interference with food, physiology and habitats of various organisms, and reduced capacity of ecosystems to sustain themselves. These implications have negative socio-economic impacts and as such, the management of WW should become a priority.

1.1 Introduction

Water is a vital natural resource whose demand in recent times has grown exponentially. A report by the UN-Water (2021) detailed a six-fold increase in freshwater use globally in the past century with at least 1% increase in demand for the commodity since the 1980s. The rise in water consumption is closely associated with deterioration of its quality (Lin et al. 2022). Contemporary society is characterized by expanded urbanization, population rise, concerted efforts to improve agricultural production and a rise in industrialization, which have resulted to pollution and degradation of the environment including water resources that are essential for life (Nyika and Dinka 2022a; Nyika 2022). The prevailing conditions are driven by production of wastewater (WW) that has the potential to derail the realization of sustainable development goals (SDGs) especially when the sewage from both domestic and industrial sources is discharged to the environment without adequate treatment (Xu et al. 2022). The affected SDGs include SDG 1, 2, 3, 6, 7, 8, 9, 11, 12, 13, 14 on zero poverty;

zero hunger; good health and wellbeing; clean water and sanitation; affordable and clean energy; decent work and economic growth; industry, innovation and infrastructure; sustainable cities and communities; responsible consumption and production; climate action; life below water, respectively (Obaideen et al. 2022).

A study by Qadir et al. (2020), noted that more than 380 billion m^3 of WW was produced globally following the aggregation of available country-specific data. The reported volume was equated to nearly five times the mean yearly volume of the Niagara Falls whose flow rate was estimated at 2407 m^3 per second (World Waterfall Database 2017). The total volume of WW can be used to irrigate more than 31 million ha of cropland without further dilution and assuming two yearly crop growing seasons and zero water losses (Qadir et al. 2020). More than 29 million ha of cropland across the globe was directly and indirectly irrigated using untreated WW especially in urban settings (Thebo et al. 2014). Further increases in global production of WW are expected in the coming years. Qadir et al. (2020) estimated that production rates would rise by 24% from the current rate of 380 billion m^3 to 470 billion m^3 by 2030. Further estimates showed that production rates would rise by 51% of the current rate to 574 billion m^3 by 2050 (Qadir et al. 2020). These projections allude to the need to better manage and treat WW and make it an alternative source of water for non-priority water uses.

WW management and treatment is a global challenge, exacerbated by poor sanitation and an increase in anthropogenic activities. Kumari et al. (2023) noted that the situation is dire in underdeveloped nations where barely 10% of all the produced WW is treated. Poor WW treatment is a prevailing trend despite the development of treatment and collection technologies (Zhongming et al. 2020; Onu et al. 2023). The rest of the unprocessed and untreated WW mainly sourced from kitchens, toilets, household chores, other pharmaceuticals and medicine, from dairy and tannery industries as well as agricultural waste is released to the environment or used for other purposes. According to UN-Water (2022), 45% of the generated municipal WW is released to the environment irrespective of its collection or non-collection. The release and reuse of WW becomes an environmental and public health concern, which causes extensive pollution on freshwater bodies in addition to harming aquatic life due to its contaminant and pollutant constituents including microbial pathogens, organic pollutants, plant nutrients and heavy metals (Kesari et al. 2021; Nyika 2022). Ultimately, natural and man-made resources are affected leading to economic downtrends associated with freshwater inaccessibility and poor human health and wellbeing.

The imminent increase in WW production trends and the environmental and public health concerns associated with its use without any processing and treatment necessitate further action. Kumari et al. (2023) highlighted the need for genuine management while maintaining social, political, technical and economic equilibrium in the WW sector as the basis for sustainable development. Such a management approach begins with establishing the trends in WW production, the sources and composition of the resources and the effects of using it without any form of treatment. However, data on these aspects is inconsistent, unavailable, scattered and in some cases infrequently reported and monitored (Mateo-Sagasta et al. 2015; Qadir et al. 2020; Nyika 2022). These limitations make WW management and subsequent valorization and recovery of its associated resources difficult to plan and achieve. Additionally, it derails efforts to realize SDG 6.3, whose aim is to reduce the proportion of untreated WW by 50% in 2030 through enhancing its safe reuse and recycling. Other SDGs affected by unscientific management of WW include SDG 7.1 on improved energy security and SDG 13.1 on climate change adaptation and mitigation. To bridge this gap, this chapter uses recent literature to identify trends in WW production globally, establish the sources and composition of the resource and detail the effects of its non-treatment to the environment and public health.

1.2 Sources of Wastewater

Sewer systems usually collect WW from many sources before transferring it for treatment. Some of the main sources and categories of WW are as shown in Fig. 1.1. They include domestic, industrial, agricultural and healthcare sources, landfill leachate and from storm water runoff. The specific characteristics and strengths of pollutants in the categories varies widely based on their production sources (Ahmed et al. 2017). Municipal or domestic WW is produced in institutions, from recreational facilities, commercial districts and residential areas. It comprises of liquid and solid discharges of animals and humans and its unsafe management is a threat to public health. The WW is a mixture of household wastes such as garbage, trash, detergents, household cleaners and papers which along with black and grey water end up in the sewer system (Ali et al. 2021). During their collection, storage and pretreatment, they accumulate many pathogenic microbes including bacteria and viruses.

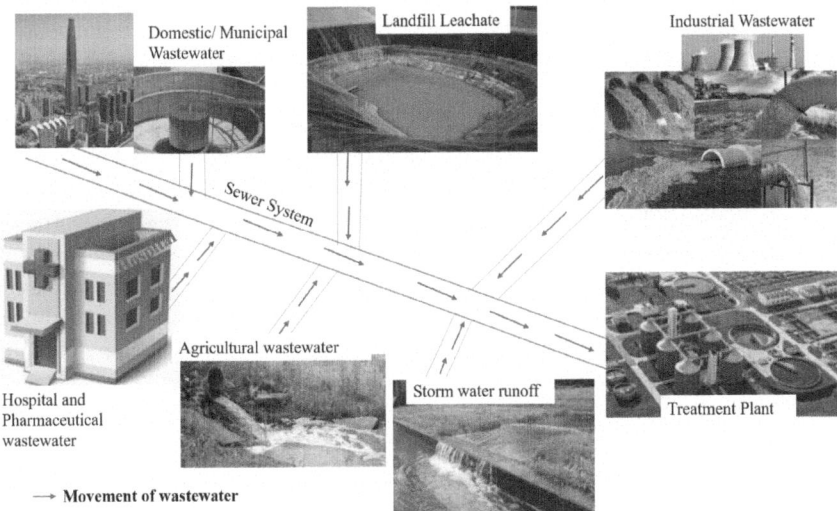

Fig. 1.1 Sources of wastewater. Drawn by the authors

Industrial WW is produced during various manufacturing processes and is released directly or after treatment to sewer collection systems from the firms. The WW contains toxic matter, detergents, grit, alkalis, dyes, acids and other chemical substances. Main polluters are industries engaging in petroleum refining, textile manufacturers, laundries, printing firms, milling and tanning industries (Ali et al. 2021).

Storm water runoff, which results from excess rainfall in paved urban areas is another source of WW directed to sewer systems. It mainly constitutes of a mixture of water with grit, gravel and sand and can be designed separately or in combination with municipal WW. During the collection of municipal and industrial WW, the conveyance pipes can leak introducing WW into the aquifer systems where groundwater becomes polluted. Additionally, the mismanagement of solid waste disposed at landfills and dumpsites can result to the formation of noxious leachate that can infiltrate into the vadose zone and saturated systems introducing pollution into natural resources (Nyika 2021; Nyika et al. 2022).

Agricultural farmlands using irrigation water along with agrochemicals, soil conditioners and fertilizers can generate WW, which has hazardous and toxic substances. Similarly, hospital processes and manufacturing activities in pharmaceutical industries generate WW containing hazards and toxic chemicals such as refractory micropollutants (Ali et al. 2021). If such WW is mixed with municipal WW, it becomes increasingly difficult to treat using conventional technologies.

The sources of WW can also be broadly grouped into two (point or non-point) based on the nature of their introduction to the environment (Ali et al. 2021). Municipal sewage and industrial WW are categorized as point sources since they are collected using a network of channels or networks and afterwards, conveyed to a common source of discharge area. Municipal sewage is collected from offices, schools, homes, stores and buildings and includes additional industrial discharge, which is allowed to be collected to a sanitary collection system. Such point sources of pollution can be managed by WW and waste avoidance and its scientific treatment before discharge to freshwater resources (Figoli et al. 2017).

Storm water runoff, agricultural WW, hospital and pharmaceutical WW are classified as non-point sources because they have many discharge points. In these sources treatment at each outlet is difficult because the WW flows along natural drainage channels on the land surface to nearby water bodies and collection and transportation using pipes or channels only occurs for short distances to a discharge point (Ali et al. 2021). The management of non-point sources of WW is complex and requires huge infrastructural investments. Potential management approaches involve the modification of land-use practices, improved education to farmers, manufacturers and city planners and improved engineering works on sewer lines to reduce WW generation and eliminate combined sewer overflow systems that complicate WW treatment (Ali et al. 2019, 2021).

1.3 Global Trends on Wastewater Production and Management

Several studies focusing on estimating the amount of WW generated globally have been conducted and resulted to different findings. A study by Sato et al. (2013) using published and electronic data estimated the yearly WW generated in 2013 to be 330 billion m^3. Similarly, Qadir et al. (2020) estimated that WW produced in 2019 was 380 billion m^3. Jones et al. (2022) estimated that annual WW production levels would rise to 490 billion m^3/year, which showed a 42 and 50% increase from the values estimated in 2013 and 2015, respectively.

The mean per capita production of WW was approximately 49 m^3 annually (Jones et al. 2021). The value was lower compared to estimates by Qadir et al. (2020) whereby global WW per capita of 2019 was 95 m^3/capita/year. In specific regions, the yearly volumes of WW generated in 2019 was estimated at 46, 82, 65, 114, 88, 124 and 231 m^3/capita/year in Sub-Saharan Africa (SSA), Asia, Latin America and Caribbean, Middle East and North America, Oceania, Europe and North America, respectively (Qadir et al. 2020). Regions such as the United States of America (USA), Canada, prosperous but small European nations, Australia and Monaco had high WW per capita at 211, 198, 257, 220 and 203 m^3 annually in respective order. Larger Western Europe countries including the United Kingdom, France, Germany and Northern Ireland and Germany had lower production levels between 66 and 92 m^3 per year. The SSA region produced less WW at 10 m^3 per capita yearly (Jones et al. 2021). Figure 1.2a showed the estimated domestic and industrial WW production rates of 2015 in m^3/year per capita as reported by Jones et al. (2021).

Using different sources, global municipal WW production rates are rising and the trend is projected to continue in the near future as shown in Fig. 1.3. Using WW generation levels of 2015 (Qadir et al. 2020) municipal WW production globally will rise by 51% up until 2050. The production rates will be more pronounced in Asia, Europe and North America compared to SSA and Oceania regions (Fig. 1.3). Asia was the largest WW producer compared to other regions at 42% and expected to increase to 44% by 2030. The trend corresponds to the region's high urban population unlike other regions (Liao et al. 2021). Europe and North America were also large WW producers. However, North America had a lower per capita use of WW producing comparatively lesser resource compared to Europe that had a lesser urban population (Qadir et al. 2020). The SSA region also recorded five times lower volumes of generated WW compared to North America. The trend is associated with a flawed water supply and infrastructural system as well as poorly monitored and managed sewage collection and treatment system in the region and particularly, in urban settings (Qadir et al. 2020; Nyika et al. 2022; Onu et al. 2023).

Reports on WW treatment are largely subjective in that they estimate that more than 80% of the generated WW is released to the environment without treatment or reuse (UN-Water 2018; World Bank Group 2020; Petrik et al. 2022). This figure is difficult to interpret considering that SDG indicator 6.3 only quantifies the proportion of industrial and domestic WW flow that is safely treated. Although most reported

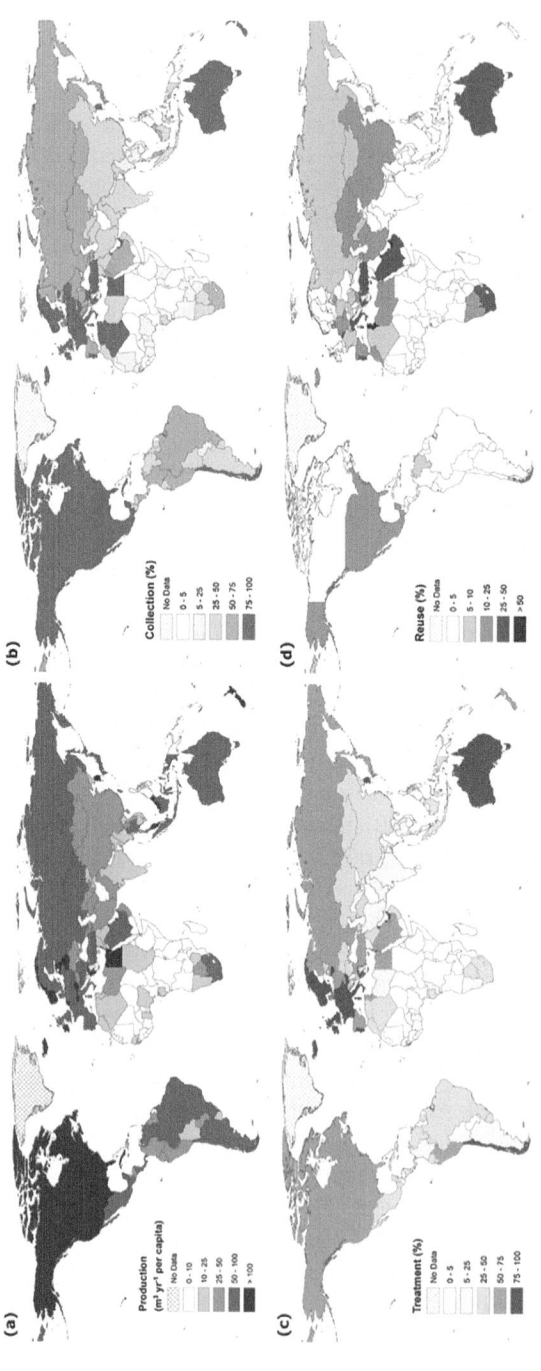

Fig. 1.2 Global estimates of wastewater showing **a** production rates in m³/year per capita for the domestic and industrial sector, **b** percentage share of collected WW, **c** percentage share of treated WW and **d** percentage share of reused water as of 2015. Reproduced from UNEP (2023)

1.3 Global Trends on Wastewater Production and Management

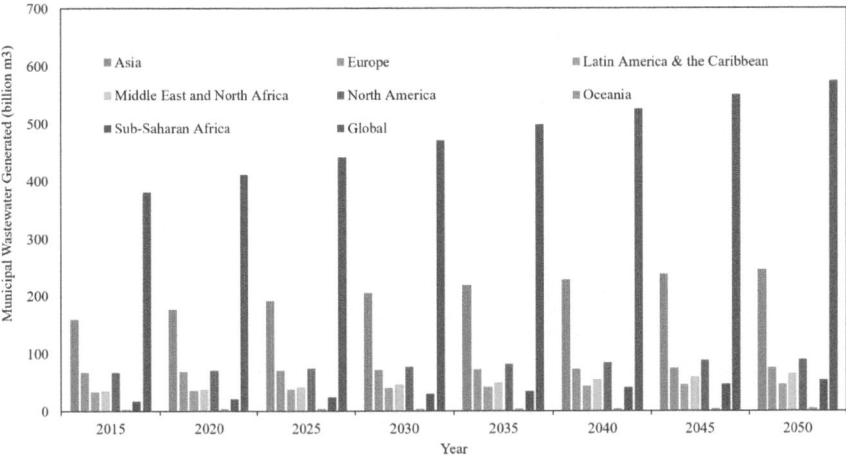

Fig. 1.3 Predicted levels of municipal wastewater generation across the globe between 2015 and 2050. Drawn by authors and adopted from Qadir et al. (2020)

data is only on domestic WW treatment in urban areas, while rural sewage production is neglected. In a documented report on the progress of the indicator in 2021, only 14 SSA countries of the possible 48 included industrial WW in the amounts generated (UN-Habitat and WHO 2021). Furthermore, at least 60% of the resource was reported as untreated. The trend alludes to data paucity and inconsistencies on the global levels of industrial WW treatment.

For urban and rural domestic WW and in reference to the SDG indicator 6.3.1, approximately 50% is released to the environment without any form of treatment (UN-Habitat and WHO 2021). Although the conclusion was made using data collected from Organization for Economic Co-operation and Development (OECD), Eurostat and UN-Statistics Division (UNSD) databases, the report discredits the quality of the data and its incomprehensiveness. Due to data rarity to compile the report, only 22% of the population in 56 countries across the globe was used. The nations generated more than 131,871 million m^3 of both municipal and industrial WW in 2015 and out of the total, 32% was treated (UN-Habitat and WHO 2021). Only 32 of the 56 countries provided data on industrial WW generation while only 15 nations reported on the treatment of the resource. The reported trends suggest that WW monitoring is not prioritized globally.

Jones et al. (2021) estimated that 63% of municipal WW in urban settings was collected while 51 and 11% of the resource was treated and reused, respectively (Fig. 1.2b, c). Europe and North America recorded the highest levels of collection and treatment compared SSA region. The trend is associated with economic development and the financial, technological and human capacity to manage WW and its infrastructure in the developed countries compared to the global south. However, the reported data is a result of modelling and has uncertainties and discrepancies in the regions based on their research capacity and economic development and can be improved through improved monitoring of WW services in both rural and urban settings globally (Qadir et al. 2020; Jones et al. 2021, 2022; Nyika 2022; UNEP 2023).

An even smaller portion of the water is directed for reuse after treatment. According to Jones et al. (2021), only 11% of the manufacturing and domestic WW produced is channeled back for reuse. Figure 1.2d shows the global levels of WW reuse. Although with data gaps, the percentage reuse of the resource is low compared to the untapped potential estimated at 320 billion m^3 annually and with a ten times capacity to supply water compared to the world's desalination capacity (Jones et al. 2021). Taking up WW reuse could therefore ease the existing global pressure of water scarcity. For this reason, large producers such as the USA, China, India and Indonesia have advanced to enhance their WW treatment and reuse capacities by 16, 40, 15 and 12 billion m^3/year, respectively in order to realize SDG 6.3 by 2030 (Jones et al. 2022).

Western Europe, the Middle East and North Africa were the best WW re-users (Fig. 1.2d). A contrast trend was evident in SSA and South Asia regions due to the preference to use conventional water supplies and apparent limitations in WW collection and treatment (UNEP 2023). Strong regulatory measures at country level could greatly enhance WW reuse. This trend is evident in Israel and Singapore whereby more than 25% and about 40% of the respective nations' water demand is met using reused WW (Kehrein et al. 2020). China is also making concerted efforts to increase its WW reuse from 10 to 15% in 2019 to more than 30% as Xu et al. (2020) highlighted. A similar trend should be taken up especially in countries of the global south where estimated levels of WW reuse were low.

Evidently, data on WW generation, collection, treatment and reuse trend has gaps leading to inconsistencies in establishing patterns. This is particularly so when monitoring industrial and fecal sludge containing WW. According to the UN-Habitat (2023) report, the data gaps hinder WW service delivery, the improvement of its infrastructure and its regulation on its discharge. Other authors pointed similar data inconsistencies in the WW management sector pointing it as a hindrance to coherent management of the resource and inability to realize SDG 6.3 (Qadir et al. 2020; Jones et al. 2021, 2022; Onu et al. 2023; UNEP 2023). As such, concerted efforts in monitoring, collecting, sharing and aggregating WW data globally should be prioritized as strategies to increase monitoring and management capacity and develop a policy framework to regulate its discharge with minimal effects on the environment and public health (UNEP 2023; Drechsel et al. 2024).

1.4 Wastewater Characteristics and Composition

1.4.1 Wastewater Characteristics

The constituents and characteristics of WW has changed over time due to industrial revolution. Prior to the industrial era, most municipal WW emanated from domestic activities but with advanced technology, industrial discharge has increased exponentially (Singh et al. 2023). The resultant WW comprises of organic compounds and inorganic substances most of which are of toxic nature. Depending on the source of WW, the concentrations and amounts of the specific substances differ. The concentrations of the substances are influenced by regional, seasonal and diurnal fluctuations and can increase or decrease in amounts based on socio-economic conditions of generators, weather patterns and lifestyle changes in targeted production regions (Trianni et al. 2021). The substances can be classified as having physical, chemical or biological characteristics once analysis of WW is done. The WW indicators of the three classes of characteristics and their associated significances in management aspects of the resource are as summarized in Table 1.1.

1.4.2 Composition of Wastewater

WW from different sources (Fig. 1.3) has different contents. Municipal contains household WW, yellow water, kitchen water, grey and black water. In this context, black water contains fecal matter while yellow water has high levels of iron because of rust contamination and grey water is from households and offices but does not contain fecal matter. In addition, municipal WW contains industrial wastes that require special pretreatment and treatment precautions to ensure they comply with predefined standards (Ali et al. 2021). Table 1.2 shows the composition of WW and the associated concentrations of the constituents.

Industrial WW contains effluents from printing houses, photolabs, galvanic, electronic, petroleum refining, textile, pharmaceutical, milling and tanning industries. Dairy factories, breweries, slaughterhouses, commercial laundries and vegetable canneries also produce industrial WW. Effluents from industries that are released to the environment without any form of treatment are a growing concern in contemporary society due to their hazardous components and ability to pollute natural resources (Singh et al. 2023; Kato and Kansha 2024). Table 1.3 shows some of the pollutants found in industrial WW from different sources and their respective levels based on data availability.

Underground water, surface runoff and rainfall can enter in a landfill facility and mix with the solid waste during consolidation and compaction resulting to leachate formation (Nyika et al. 2022). The composition of leachate depends on the nature of the solid waste disposed in the landfill, physical, chemical and biological reactions within the site and the age of the facility. Depending on the age of the landfill, old and

Table 1.1 Physical, chemical and biological characteristics of WW and their significance to the management approaches of the resource (Ali et al. 2021)

WW indicator	Significance
Physical characteristics	
Total fixed dissolved solids (FDS)	– They characteristics are related to the viscosity of WW and influence it transport, movement, settleability and ability to be dewatered. These aspects are determinants of the potential to reuse the resource and inform on the best treatment technologies to use
Volatile dissolved solids (VDS)	
Total dissolved solids (TDS)	
Fixed suspended solids (FSS)	
Volatile suspended solids (VSS)	
Total suspended solids (TSS)	
Total fixed solids (TFS)	
Total volatile solids (TVS)	
Total solids (TS)	
Turbidity (Nephelometric turbidity units, NTU)	– Indicates the suspended particles in WW and assesses its quality and clarity after treatment
Particle size distribution (PSD)	– Informs on the settleability of solids, which determines the energy to be used during dewatering and flocculation
Settleable solids	– Shows the solids that can settle due to gravity in a given period
Conductivity	– Assesses the suitability of treated WW for reuse in agricultural irrigation
Density	
Temperature	– A key consideration in the design and operation of biological treatment facilities
Odors	– An indicator of biological activity and the presence of pathogenic microorganisms in WW
Transmittance as %	– To assess the suitability of treating sewage using UV disinfection technology
Color	– Determine if the WW is fresh or septic
Organic chemical characteristics	
5-day carbonaceous biochemical oxygen demand ($CBOD_5$)	– Determines the oxygen requirement for biological stabilization of WW
Ultimate BOD (UBOD)	
Total organic carbon (TOC)	– It is used as a substitute for BOD
Chemical oxygen demand (COD)	
Nitrogenous oxygen demand (NOD)	– Determines the amount of oxygen required to biologically oxidize WW nitrogen to nitrates
Specific organic compounds	– To determine their presence and device treatment measures to remove them from WW
Inorganic chemical characteristics	
Gases such as O_2, H_2S, NH_3, CO_2, CH_4	– Determine their presence or absence in sewage

(continued)

1.4 Wastewater Characteristics and Composition

Table 1.1 (continued)

WW indicator	Significance
Specific inorganic elements and compounds	
Metals and metalloids including As, Ca, Cd, Co, Cr, Cu, Hg, Mg, Mo, Na, Ni, Pb, Se, Zn	– To assess the toxicity of WW, its reuse potential and possibility of applying biological treatment – To determine the recovery potential for useful metals
Sulfate (SO_4^{2-})	– Associated with odour in WW
Chloride (Cl^-)	– Assess the suitability of WW for agricultural reuse
Alkalinity	– Quantifies the buffering capacity of effluent
pH	– Determines the hydrogen ions found in aqueous WW
Organic phosphorous (Org P)	– The indicators determine the nutrient content of WW, its decomposition degrees and oxidation extent
Inorganic phosphorous	
Total nitrogen (TN)	
Nitrates (NO_3^-)	
Nitrites (NO_2^-)	
Total Kjeldahl nitrogen (TKN)	
Organic nitrogen (Org N)	
Free ammonia (NH_4)	
Biological characteristics	
Coliform organisms	– Determine the presence of pathogenic microbes such as bacteria and the suitability of treating the water through disinfection
Specific pathogens including viruses, helminths, protozoa and bacteria	– Assesses presences of specific organisms to determine the suitability of WW for reuse
Toxicity	– Determine if the WW inflicts acute or chronic toxicity – Provides information on the lifestyle, diet and health of a community producing WW
Antibiotics such as sulfamethoxazole, penicillin, cloxacillin, chlortetracycline, ampicillin and amoxicillin	– Inform on the treatment technology to apply since some antibiotics reduce WW treatment efficacy by attacking nitrifying bacteria and activated sludge – Enable modifications on the design and operations of treatment plants to enhance efficiency
Antibiotic resistant bacteria such as *Enterococcus hirae* and *E. faecalis*	
Antibiotic resistant genes including tetracycline (TET) and sulfonamide (SUL) resistant genes	

Table 1.2 Some of the pollutants in municipal WW and their concentrations (Ali et al. 2021)

Pollutant (units)	Municipal WW	Household WW	Storm water	Kitchen WW	Grey water	Yellow water
Temperature (°C)	10–21	24				
TS (mg/l)	350–120	252–3320	112–1894	4101	750	
SS (mg/l)	140–450	22–1690	170–515			
VSS (mg/l)	80–320					
TDS (mg/l)	250–950	150–380	20–108			
pH	7–8	6–8.4	6–8.6	9.8	7.0–7.4	9.3
Alkalinity (equivalent/m^3)	3–7	6.3–19	5.9–92	11.4	1.4–6.4	
Conductivity ($\mu\Omega$/cm)	700–1200	600–800	29–266	1510	226–820	27
Chloride (mg/l)	500	20–50	976			
Sulfate (mg/l)	50		0.11–169			
Bacteria/protozoa (no/100 ml)	10–10					
BOD (mg/l)	150–530	112–1101	7–56	1850	120	3500
COD (mg/l)	210–740	139–1650	82–178	8071	537	7800
VOCs (mg/l)	100–440					
Total nitrogen (mg/l)	20–80	44–189	3.5	25.9	2.4	4000
TKN (mg/l)	20–80	16–189	1.9–4.18			6600–6900
NH$_3$ (mg/l)	12–50	1.6–94	5–56		< 0.05–1.3	< 0.1
NO$_3$ (mg/l)	0.5	20–40	0.86–2.21	0.96	< 0.05–0.8	< 0.1
Total P (mg/l)	4–14	0.2–32	0.35	4.3	1.2	350
Amoxicillin/ampicillin (μg/l)	150					
Tetracycline	1.1					
TET (copies/ml)	$10^{7.2}$–10^9					
SUL (copies/ml)	$10^{5.46}$–$10^{7.54}$					
Arsenic (μg/l)		1–5		1–50.5		
Cadmium (μg/l)	1–4		0.1–14		< 1	

(continued)

Table 1.2 (continued)

Pollutant (units)	Municipal WW	Household WW	Storm water	Kitchen WW	Grey water	Yellow water
Mercury (μg/l)	1–3		0.6–1.2		0.44	
Lead (μg/l)	25–80		182–443		< 10	
Iron	400–500					

fresh leachate has different concentrations of pollutants as shown in Table 1.4. The differences in composition is because fresh leachate is undergoing aerobic decomposition and mainly comprises of biodegradable organic matter unlike the old leachate, which has stabilized, is past the decomposition phase and is less acidic (Lindamulla et al. 2022).

Hospital and pharmaceutical manufacturing WW contains endocrine disrupting chemicals (EDCs), volatile organic compounds, personal care products (PPPs) and other hazardous substances. It consists of WW from surgical procedures, laundries, laboratory analyses, X-ray departments, use of glassware and catering operations at community hospitals and various care providers. Some of the pollutants found in this class of WW are as shown in Table 1.5.

Agricultural WW mainly emanates from cleaning operations during food processing and transportation (Ali et al. 2021). Dairy processing companies, breweries, slaughter houses, canning of fruits and vegetables, grain milling and sugar confectionery processes produce such effluents. The quality and quantity of WW varies based on the foods being processed and the technology applied. The misuse and overuse of agrochemicals and chemical fertilizers also contributes to agricultural WW (Srivastav 2020). Agricultural WW unlike storm water, which collects soil, sediments, and trash as it moves through sewer lines, contains EDCs, pesticides and polycyclic aromatic hydrocarbons.

1.5 Management of Wastewater

As established, WW sources are diverse in nature and so is its composition. Trends on its management show that only a small percentage of the generated WW is collected and treated prior to release to the environment. Direct release and reuse of effluent for irrigation results to negative implications due to contents such as pathogenic microbes, EDCs and heavy metals that interact with crops and soils (Kesari et al. 2021). The pollutants from WW bioaccumulate in plants and in water bodies and eventually end up in aquatic resources (Agoro et al. 2020). WW also interferes with natural salinity levels of the soils reducing its inability to grow crops. In a study by Nyika and Dinka (2022b) vegetables and soils irrigated with untreated effluents in suburban regions of Nairobi, Kenya were found to be enriched with

Table 1.3 Common pollutants in industrial WW and their concentrations (Ali et al. 2021)

Pollutant	Milling and tanning industries	Mining industry	Textile industry	Petroleum refining	Commercial laundries	Galvanic industries	Electric industry	Photolabs	Printing house
TS		1–15,878	1560						
SS			156	441	1000				1180
TDS		120–8870							
pH		2.67–7.7	9.4						
As	6.0–22.0		0.0–0.48	0.0–1.6					
Cd		400–1000							
Cr	30–70			0.6	0.88				0.8–1.0
Cu				0.0–1.4	1.7		100–200		
Pb	0.02–4.6	8.4		4.5		0–5			
Ni	0.01–5.0			0.29					
Ag		5–130		7–120				230	1.0–1.3
BOD	1000–2000		6000	0–5000	1300			400–700	210
NH$_3$	83–159	0.53–22							
Fats, oils, grease	40–200		21–303		8–795				

*All the parameter units are in mg/l

1.6 Effects of Poor Wastewater Management

Table 1.4 Pollutants found in old and fresh leachate and their respective concentrations (Ali et al. 2021)

Pollutant levels (mg/l)	Old leachate	Fresh leachate
SS		200–2000
pH	7–8	4.5–5.5
Magnesium		50–1000
Calcium		200–3000
Chloride		200–3000
Sodium		200–2000
Iron		50–1000
COD	500–3000	3000–60,000
BOD	50–200	2000–30,000
TOC	100–1000	1000–20,000
Volatile fatty acids	50–100	1000–25,000
Ammonia		95–475
Total P		5–100
Total N		20–1000
Cr		50–600
Cd		1–10
Fe		50–600

heavy metals. A similar trend was established by Hussain et al. (2019) in India. Once consumed by humans and animals, the crops result to heavy metal bioaccumulation and ultimately, health risks. Elevated levels of arsenic lead to bone and kidney cancers while Cd results to musculoskeletal complications (Kesari et al. 2021) Mercury interferes with the function of the nervous system and copper poisoning results to anemia, nausea, vomiting, kidney and liver damage (Kesari et al. 2021). The following sections discuss the health, environmental and socio-economic impacts of releasing and reusing untreated effluents.

1.6 Effects of Poor Wastewater Management

1.6.1 Health Impacts

(a) **Disease Burden**

Poor management of WW and fecal sludge has devastating effects on human health associated with inadequate sanitation. Pathogenic microbes and organic material mainly found in municipal and household WW enables the negative human health effects. In low- and middle-income countries, the trend has been associated with diarrhea, which is a main cause of disease and death among children below five

Table 1.5 Composition and levels of substances found in WW from hospitals and pharmaceutical industries (Ali et al. 2021)

Pollutant	Units	Concentration
SS	Mg/l	120–400
Chloride		65–360
NH_4		10–68
Total N		5–80
DOC		120–130
TOC		31–180
COD		1350–2480
BOD		200
NO_2		0.1–0.58
NO_3		1–2
Total surfactants		4–8
Detergents		4.9
Fats, oils, grease		50–210
Phosphate		6–19
Total P		0.2–13
Conductivity	μS/cm	300–1000
pH		7.7–8.1
Total coliforms	MPN/100 ml	10^5–10^8
Fecal coliforms		10^3–10^7
E. coli		10^3–10^6
Cytostatic agents	μg/l	5–50
Lipid regulators		1–10
Antiinflammatories		5–1500
Total disinfectants		2–200
Mercury		1.65
Platinum		13
Antibiotics		30–200
Cytostatics		24
Gadolinium		32
Adsorbable organic halogens (AOX)		550–10,000
Intensive care medicine (ICM)		0.2–2600
Hormones		0.16
β-blockers		0.4–25

years (WHO and UNICEF 2021). Other waterborne diseases associated with exposure to untreated WW include cholera, schistosomiasis, polio, trachoma, malabsorption, typhoid and dysentery (Kesari et al. 2021; WHO and UNICEF 2021). The diseases are spread by microbial species including *E. coli, Legionella spp., Vibrio cholera, Salmonella typhimurium, Pseudomonas aeruginosa, Giardia intestinales* and *Shigella sonnei* among others (Kesari et al. 2021).

Infections also result from exposure to WW and subsequent sanitation inadequacy. In many children, the infections lead to stunted growth, which is often a result from diarrhea, malnutrition, environmental enteric dysfunctions (EED) due to pathogenic infections and ultimately, poor cognitive and physical development (WHO 2018; Bhave et al. 2020; UNESCO 2020; Root 2022).

Health conditions and diseases around the world associated with water, sanitation and hygiene (WASH) result to more than 1.6 million fatalities annually and at least 105 million disability adjusted life years (DALYs) (UN-Habitat 2023). More than 820,000 of the total deaths accounting for nearly 60% share are a result of diarrheal diseases while the rest are a result of acute respiratory infections, schistosomiasis and protein-energy malnutrition burden. Contact with WW contaminated soil and water is also associated with helminthic infections and trachoma (Pruss-Ustun et al. 2019).

Diseases associated with untreated WW mainly affect vulnerable groups especially in poor developing countries. Migrant communities and urban informal settlers are such vulnerable groups with limited access to good quality water, WW collection and treatment services, mostly practice open defecation and do not have adequate sanitation and hygiene facilities. Communities living downstream of WW outlets and in low-lying areas are at risk of exposure from pathogens emanating from unmanaged effluents. Benova et al. (2014) also highlighted that management of WW in hospitals and healthcare centers is imperative in reducing the risk of waterborne infections among infants, elderly and pregnant mothers, which can result to sepsis and mortalities.

Sanitation workers handling WW such as manual pit emptiers are also vulnerable to its associated diseases. Their etiology to respiratory dysfunctions and gastroenteritis is high according to a meta-analysis on their health impacts (Oza et al. 2022). The workers are also exposed to toxic gases, effluents and sludge, which can result to skin burns, eye infections, schistosomiasis, cryptosporidiosis, polio, hepatitis, typhoid and cholera (World Bank, ILO, WaterAid and WHO 2019). Once exposed to fecal sludge containing noxious gases such as ammonia, carbon dioxide, nitrogen, methane and hydrogen sulfide, the workers can inhale them leading to fatalities (WHO 2018).

The diverse nature of pollutants found in untreated WW pose as health risks to consumers, farmers and communities if used to irrigate crops. Harmful antibiotic residues have been isolated from WW irrigated soils and food-borne and diarrheal diseases have been related to the practice (Adegoke et al. 2018). Fungal infections and dermatitis were also associated with exposure to untreated agricultural WW in Vietnam (UN-Habitat 2023). Helminthic infections and heavy metal poisoning have also been reported following the use of untreated sewage in crop irrigation (Kesari et al. 2021).

(b) **Exposure to Chemicals**

The constituents of WW from the health sector, industrial, agricultural and municipal sources contains chemicals that are disease causing. Such chemicals include heavy metals, EDCs, pesticides, volatile compounds, hydrocarbons and organic materials. The chemicals are intractable, persistent in human bodies, bioaccumulative in nature, and non-biodegradable and are associated with eye irritations, lung infections, skin cancer and chromin dermatoses (Kesari et al. 2021). WW containing pharmaceuticals and PPPs contains EDCs such as ethinylestradiol, estradiol, estrone and nonylphenol, which are highly toxic even in low concentrations. The EDCs are also found in textile, distillery, tannery and paper industry WW (Haq and Raj 2019). EDCs have been associated with breast cancer, polycystic ovarian syndrome, reduced semen quality, low birthweight, gestational diabetes, impaired glucose intolerance, obesity and prostrate cancer among other health complications (Kahn et al. 2020). WW also contains perfluorinated compounds (PFCs) that are associated with carcinogenesis, neurotoxicity, immunotoxicity, reduced fertility, birth defects and pre-eclampsia (Lee 2018).

Nitrogenous compounds found in WW contaminate drinking water and are toxic if ingested. Potential health complications include neural tube defects, thyroid disease, colorectal cancer, methemoglobinemia in children and reproductive problems (Ward et al. 2018; UN-HABITAT 2023). Sewage also contains mucroplastics that are toxic to humans (Ormaniec 2024). During application of untreated or inadequately treated WW for agriculture, microplastics can be inhaled or ingested by humans leading to altered metabolism, carcinogenic effects, reproductive toxicity, neurotoxicity and oxidative stress (Lee et al. 2023).

1.6.2 Environmental Impacts

The release of untreated WW including black water to the environment affects natural land and water resources (Wear et al. 2021). The main environmental impacts include soil pollution, contamination of aquifer systems, pollution of surface water and negative effects on biodiversity (Thomas et al. 2018). The presence of heavy metals, suspended solids, EDCs, pathogens and nutrients in sewage impact natural habitats once introduced in the environment (Wear et al. 2021).

In coastal ecosystems, introduction of untreated WW leads to a plethora of effects including fisheries decline due to chemical toxicity, eutrophication due to nutrient enrichment of sea water, loss of habitats and degradation (Wear and Thurber 2015). A study by Tuholske et al. (2021) established that 88% of seagrass hotspots in coasts of China, Nigeria, India and Ghana as well as 58% of coral hotspots in coastal Yemen, India, Haiti, Kenya and China are enriched with nitrogen from WW input. Wear et al. (2021) noted that the introduction of WW in coastal water bodies resulted to algae

1.6 Effects of Poor Wastewater Management

overgrowth, expansion of dead zones, the bleaching of coral reefs and their inhibited growth and reproduction and ultimately, death. Oyster reefs, which play an important role in filtering toxins are on a decline due to WW release in coastal bodies. Wear and Thurber (2015) reported that reproductive capabilities and the survival of oyster reefs was at threat due to effluent pollution of seas and oceans. Microbial communities in coastal China resulted to reduction in the species in downstream regions and their reduced diversity due to phylogenetic clustering (Dai et al. 2023).

Freshwater resources are vulnerable to WW discharge since they are located in close proximity to human settlements. Surface water bodies of both developed and developing nations suffer from the discharge of the effluents. For this reason, the most polluted rivers in populous countries such as India (Wear et al. 2021) and China (Tuholske et al. 2021) are found in the vicinity of human settlements. Many African rivers and lakes are polluted due to the release of effluents from textile industries (Hepworth et al. 2021) while 46% of streams and rivers and 35% of lakes in the USA are polluted by agricultural WW (Wear et al. 2021; UN-HABITAT 2023). Pollution in freshwater bodies is associated with diseases in shellfish and fish and the emergence of dead zones. The trend is attributable to EDCs and the oxygen depleting substances found in WW. UNESCO (2020) reported that more than 80% of plants and animals are at risk of extinction due to mismanagement of WW among other reasons. Sewage in freshwater resources also results to white pox disease in corals (Wear et al. 2021), heavy metal accumulation in predatory fish (Saha et al. 2016) and ingestion of WW constituents in juvenile salmon (Meador et al. 2016).

Unregulated release of WW to the environment also affects terrestrial ecosystems and particularly wildlife and their species diversity. In many European countries, China and the USA, effluents and sludge are discharged to marshlands, forests, estuarine and open water systems where their toxic pollutants and organic elements accumulate (Wear et al. 2021; UN-Habitat 2023). The pollutant eventually enter in food chains and hence their presence in noticeable concentrations in soil and vegetation of the vicinity as well as milk of cows that used such areas as pastures. Ultimately, the trend raises concerns on the safety of releasing and reusing untreated wastewater to the environment, human and animal health. Manzetti and van der Spoel (2015) also noted that disposal of sludge in wetlands reduced their plant quality and affected food production for insects, birds and animals that dwelled in such habitats leading to biodiversity loss and limited survival opportunities for the affected populations. Heavy metals, EDCs, antibiotics and pathogens in WW were a threat to natural habitats and wildlife and are a growing public concern due to their toxic nature. Fayomi et al. (2019) noted that antibiotics interfered with microbial communities in water and land resources, distorted their structure, function and diversity due to their persistent and bioaccumulative nature.

The processes involved in collection, transport, reuse and treatment of WW also affect the environment negatively through the release of greenhouse gases (GHGs) such as nitrous oxide, methane and carbon dioxide (UN-Habitat 2023). The gases are also produced during biodegradation of organic matter in WW and their concentration vary based on the WW handling practices that are region-based, the size of treatment plants and the technologies involved in the treatment cycle. WW at treatment plants generate approximately 5% of non-CO_2 GHGs emissions globally and projections show that the levels will increase to 22% by 2030 (Maktabifard et al. 2023). Methane produced from septic tanks storing greywater and from fecal sludge contributed to 4.7% of all global emissions of the gas in 2020 alone (Cheng et al. 2022). Biological treatment of WW to remove nutrients leads to emission of nitrous oxide (UN-Habitat 2023).

1.6.3 Socio-economic Impacts

The impacts of discharge and release of WW have a wider scope to the global socio-economy. Lixil, Oxford Economics and WaterAid (2016) approximated the global cost of poor sanitation at $182.5 billion and $222.9 billion in 2010 and 2015, which is an increase based on healthcare, mortality and access to sanitation facilities. The Asia Pacific region had the highest economic burden, a trend associated with high population in the region corresponding to higher WW generation rates (UN-Habitat 2023). Economic losses ranging from 0.3 to 3.2% gross domestic product (GDP) were associated with WW generation and its management across the globe with the SSA region carrying the greatest burden (WaterAid 2021). Conversely, the management of WW and fecal sludge is expected to reap economic benefits to the tune of $86 billion (WaterAid 2021) considering reduced healthcare burden and increased productivity of more than $1.4–1.6 trillion in respective order annually (Chaitkin et al. 2022) and after considering the larger benefits of WASH access.

The socio-economic benefits are disproportionate and evidence shows that developing countries will be disadvantaged compared to developed ones (UN-Habitat 2023). In the former, communities are burdened with unmanaged WW and fecal sludge, lack appropriate WASH and healthcare services. A report by WHO and UNICEF (2021) assessing the progress in realizing WASH established a similar disproportionate pattern with poor nations being disadvantaged.

The tourism sector is also adversely affected by WW mismanagement. In Tarawa city of Kiribati, the mismanagement of waste and WW through combined sewer systems resulted to seaside degradation making it unappealing, unclean and unsafe for swimming among tourists and hence, the decline in their numbers (van Minh and Nguyen-Viet 2011). Eutrophication impacts due to nutrient enrichment of fresh and coastal water resources from WW result to economic costs due to biodiversity loss, fishery associated economic losses, increasing demand for public health services and decreased earnings from the tourism sector (McCrackin et al. 2017). The costs of importing water from distant areas due to WW pollution is also an economic downturn to many urban regions of the world (UN-Habitat 2023).

Sanitation enhance the quality of life and human wellbeing and vice versa. Untreated WW undermines efforts to enhance sanitation and therefore, demote better quality of life (Ross et al. 2021). The effects of lack of sanitation are felt more by vulnerable groups, women and girls. Girls from the developing world miss school due to lack of WASH facilities during their menstruation while women and the vulnerable such as the old, face violence and harassment in search for clean water at distant areas due to extensive pollution of the resource (WHO and UNICEF 2021). Additionally, the dignity of disabled persons is lowered from the use of inaccessible WASH facilities such as toilets and increases their abuse risk.

Evidently, the impacts of releasing and reusing WW are grave and have a wider scope. They also have the capacity to derail efforts to realize SDGs. To counter the effects of WW mismanagement, it is imperative to monitor and quantify the impact of untreated WW holistically following its introduction to the environment. Such efforts will be counter measures to better manage the resource and its associated pollutants and ultimately, it will reduce its impacts on human health, the environment, societal wellbeing and the economy (Bertanza et al. 2022).

1.7 Conclusion

WW generation rates are on a growing trend globally due to population rise and urbanization trends. However, as established in this chapter, the management of effluents is a silent global problem characterized by inadequate management, poor connection to sewerage services, and unavailability of data on WW collection, treatment and reuse. The trends are prevalent in countries of the global south especially the SSA region and deters the implementation of effective management policies and actions on the management of the resource. Management of WW is further complicated by the complex biological, chemical and physical components of sewage, some of which are recalcitrant in nature and bioaccumulate in living organisms. The sources of effluents are also diverse and range from industrial, commercial, domestic, agricultural and healthcare activities. Effects of untreated WW are equally intricate and have been shown to affect human health by introducing noxious chemicals and pathogenic components causing diseases and toxicities; affect the sustenance of the environment and at large, influence the socio-economy of affected communities. Therefore, it is imperative to prioritize WW management as efforts to realize SDGs and nurture sustainable communities.

References

Adegoke A, Amoah I, Stenström T, Verbyla M, Mihelcic J (2018) Epidemiological evidence and health risks associated with agricultural reuse of partially treated and untreated wastewater: a review. Front Public Health 6:337. https://doi.org/10.3389/fpubh.2018.00337

Agoro M, Adeniji A, Adefisoye M, Okoh O (2020) Heavy metals in wastewater and sewage sludge from selected municipal treatment plants in Eastern Cape Province, South Africa. Water 12:2746. https://doi.org/10.3390/w12102746

Ahmed M, Zhou J, Ngo H, Guo W, Thomaidis N, Xu J (2017) Progress in the biological and chemical treatment technologies for emerging contaminant removal from wastewater: a critical review. J Hazard Mater 323:274–298. https://doi.org/10.1016/j.jhazmat.2016.04.045

Ali I, Peng C, Naz I, Lin D, Saroj D, Ali M (2019) Development and application of novel bio-magnetic membrane capsules for the removal of the cationic dye malachite green in wastewater treatment. RSC Adv 9(7):3625–3646. https://doi.org/10.1039/C8RA09275C

Ali I, Naz I, Peng C, Elsalam K, Khan Z, Islam T et al (2021) Chapter 2-sources, classifications, constituents, and available treatment technologies for various types of wastewater: an overview. Aquananotechnology 11–46. https://doi.org/10.1016/B978-0-12-821141-0.00019-7

Benova L, Cumming O, Campbell O (2014) Systematic review and meta-analysis: association between water and sanitation environment and maternal mortality. Trop Med Int Health 19(4):368–387. https://doi.org/10.1111/tmi.12275

Bertanza G, Boiocchi R, Pedrazzani R (2022) Improving the quality of wastewater treatment plant monitoring by adopting proper sampling strategies and data processing criteria. Sci Total Environ 806(3):150724. https://doi.org/10.1016/j.scitotenv.2021.150724

Bhave P, Naik S, Salunkhe S (2020) Performance evaluation of wastewater treatment plant. Water Conserv Sci Eng 5:23–29. https://doi.org/10.1007/s41101-020-00081-x

Chaitkin M, Mccormick S, Alvarez-Sala Torreano J, Amongin I, Gaya S, Hanssen O et al (2022) Estimating the cost of achieving basic water, sanitation, hygiene, and waste management services in public health-care facilities in the 46 UN designated least-developed countries: a modelling study. Lancet Glob Health 10(6):e840–e849. https://doi.org/10.1016/S2214-109X(22)00099-7

Cheng S, Long J, Evans B, Zhan Z, Li T, Chen C et al (2022) Non-negligible greenhouse gas emissions from non-sewered sanitation systems: a meta-analysis. Environ Res 212:113468. https://doi.org/10.1016/j.envres.2022.113468

Dai T, Su Z, Zeng Y, Bao Y, Zheng Y, Guo H et al (2023) Wastewater treatment plant effluent discharge decreases bacterial community diversity and network complexity in urbanized coastal sediment. Environ Poll 322:121122. https://doi.org/10.1016/j.envpol.2023.121122

Drechsel P, Bartram J, Qadir M, Medlicott K (2024) The challenge of supporting and monitoring safe wastewater use in agriculture in LMIC. NPJ Clean Water 6(7):1–3. https://doi.org/10.1038/s41545-024-00364-z

Fayomi G, Mini S, Fayomi O, Owodolu T, Ayoola A, Wusu O (2019) A mini review on the impact of sewage disposal on environment and ecosystem. IOP Conf Ser Earth Environ Sci 331(1):012040. https://doi.org/10.1088/1755-1315/331/1/012040

Figoli A, Dorraji M, Amani-Ghadim A (2017) Application of nanotechnology in drinking water purification. In: Water purification. Academic Press, pp 119–167. https://doi.org/10.1016/B978-0-12-804300-4.00004-6

Haq I, Raj A (2019) Endocrine-disrupting pollutants in industrial wastewater and their degradation and detoxification approaches. In: Bharagava R, Chowdhary P (eds) Emerging and eco-friendly approaches for waste management. Springer, Singapore. https://doi.org/10.1007/978-981-10-8669-4_7

Hepworth N, Narte R, Samuel E, Neumand S (2021) How fair is fashion's water footprint? Water witness international. https://waterwitness.org/newsevents/2021/7/12/how-fair-is-fashions-water-footprint. Accessed 31 Oct 2024

Hussain A, Priyadarshi M, Dubey S (2019) Experimental study on accumulation of heavy metals in vegetables irrigated with treated wastewater. Appl Water Sci 9:122. https://doi.org/10.1007/s13201-019-0999-4

Jones E, Van Vliet M, Qadir M, Bierkens M (2021) Country-level and gridded estimates of wastewater production, collection, treatment and reuse. Earth Syst Sci Data 13(2):237–254. https://doi.org/10.5194/essd-13-237-2021

Jones E, Bierkens M, Wanders N, Sutanudjaja E, Van Beek L, Van Vliet M (2022) Current wastewater treatment targets are insufficient to protect surface water quality. Commun Earth Environ 3(221). https://doi.org/10.1038/s43247-022-00554-y

Kahn L, Philippat C, Nakayama S, Slama R, Trasande L (2020) Endocrine-disrupting chemicals: implications for human health. Lancet Glob Health 8(8):703–718. https://doi.org/10.1016/S2213-8587(20)30129-7

References

Kato S, Kansha Y (2024) Comprehensive review of industrial wastewater treatment techniques. Environ Sci Pollut Res 31:51064–51097 (2024). https://doi.org/10.1007/s11356-024-34584-0

Kehrein P, Van Loosdrecht M, Osseweijer P, Garfí M, Dewulf J, Posada J (2020) A critical review of resource recovery from municipal wastewater treatment plants—market supply potentials, technologies and bottlenecks. Environ Sci Water Res Technol 6(4):877–910. https://doi.org/10.1039/c9ew00905a

Kesari K, Soni R, Jamal Q, Tripathi P, Lal J, Jha N et al (2021) Wastewater treatment and reuse: a review of its applications and health implications. Water Air Soil Pollut 232:208. https://doi.org/10.1007/s11270-021-05154-8

Kumari S, Dwivedi S, Khan M, Nayanam S, Dhasmana A, Malik S (2023) The challenges of wastewater and wastewater management. In: Shah M (ed) Advanced and innovative approaches of environmental biotechnology in industrial wastewater treatment. Springer, Singapore. https://doi.org/10.1007/978-981-99-2598-8_5

Lee Y (2018) Potential health effects of emerging environmental contaminants perfluoroakyl compounds Yeungnam Univ J Med 35(2):156–164. https://doi.org/10.12701/yujm.2018.35.2.156

Lee Y, Cho J, Sohn J, Kim C (2023) Health effects of microplastic exposures: current issues and perspectives in South Korea. Yonsei Med J 64(5):301–308. https://doi.org/10.3349/ymj.2023.0048

Liao Z, Chen Z, Xu A, Gao Q, Song K, Liu J, Hu H (2021) Wastewater treatment and reuse situations and influential factors in major Asian countries. J Environ Manage 282:111976. https://doi.org/10.1016/j.jenvman.2021.111976

Lin L, Yang H, Xu X (2022) Effects of water pollution on human health and disease heterogeneity: a review. Front Environ Eng 2:1149950. https://doi.org/10.3389/fenvs.2022.880246

Lindamulla L, Nanayakkara N, Othman M, Jinadasa S, Herath G, Jegatheesan V (2022) Municipal solid waste landfill leachate characteristics and their treatment options in tropical countries. Curr Pollut Rep 8:273–287. https://doi.org/10.1007/s40726-022-00222-x

Lixil, WaterAid and Oxford Economics (2016) The true cost of poor sanitation. https://www.lixil.com/en/sustainability/pdf/the_true_cost_of_poor_sanitation_e.pdf. Accessed 31 Oct 2024

Maktabifard M, Hazmi H, Szulc P, Mousavazadegan M, Xu X, Zaborowska E et al (2023) Net zero carbon condition in wastewater treatment plants: a systematic review of mitigation strategies and challenges. Renew Sustain Energy Rev 185:113638. https://doi.org/10.1016/j.rser.2023.113638

Mateo-Sagasta J, Raschid-Sally L, Thebo A (2015) Global wastewater and sludge production, treatment and use. In: Drechsel P, Qadir M, Wichelns D (eds) Wastewater: economic asset in an urbanizing world. Springer Netherlands, Dordrecht, pp 15–38. https://doi.org/10.1007/978-94-017-9545-6_2

Manzetti S, van der Spoel D (2015) Impact of sludge deposition on biodiversity. Ecotoxicology 24:1799–1814. https://doi.org/10.1007/s10646-015-1530-9

McCrackin M, Jones H, Jones P, Mateos D (2017) Recovery of lakes and coastal marine ecosystems from eutrophication: a global meta-analysis. L&O 62(2):507–518. https://doi.org/10.1002/lno.10441

Meador J, Yeh A, Young G, Gallagher E (2016) Contaminants of emerging concern in a large temperate estuary. Environ Pollut 213:254–267. https://doi.org/10.1016/j.envpol.2016.01.088

Nyika J (2021) Application of experimental and modelling techniques to estimate the effects of landfill leachate on soil and water. PhD Thesis, University of South Africa

Nyika J (2022) Wastewater for agricultural production, benefits, risks, and limitations. In: Chatoui H, Merzouki M, Moummou H, Tilaoui M, Saadaoui N, Brhich A (eds) Nutrition and human health. Springer, Cham. https://doi.org/10.1007/978-3-030-93971-7_6

Nyika J, Dinka, M (2022a) A mini-review on wastewater treatment through bioremediation towards enhanced field applications of the technology. AIMS Environ Sci 9(4):403–431. https://doi.org/10.3934/environsci.2022025

Nyika J, Dinka M (2022b) Heavy metal pollution in soils and vegetables from suburban regions of Nairobi, Kenya and their community health implications. Pollution 8(4):1434–1447. https://doi.org/10.22059/POLL.2022.341522.1440

Nyika J, Dinka M, Onyari E (2022) Effects of landfill leachate on groundwater and its suitability for use. Mater Today 57(2):958–963. https://doi.org/10.1016/j.matpr.2022.03.239

Obaideen K, Shehata N, Sayed E, Abdelkareem M, Mahmoud M, Olabi A (2022) The role of wastewater treatment in achieving sustainable development goals (SDGs) and sustainability guideline. Energy Nexus 7:100112. https://doi.org/10.1016/j.nexus.2022.100112

Onu M, Ayeleru O, Oboirien B, Olubambi P (2023) Challenges of wastewater generation and management in sub-Saharan Africa: a review. Environ Chall 11:100686. https://doi.org/10.1016/j.envc.2023.100686

Ormaniec P (2024) Occurrence and analysis of microplastics in municipal wastewater, Poland. Environ Sci Pollut Res 31:49646–49655. https://doi.org/10.1007/s11356-024-34488-z

Oza H, Lee M, Boisson S, Pega F, Medlicott K, Clasen T (2022) Occupational health outcomes among sanitation workers: a systematic review and meta-analysis. Int J Hyg Environ Health 240:113907. https://doi.org/10.1016/j.ijheh.2021.113907

Petrik L, Ngo H, Varjani S, Osseweijer P, Xevgenos D, Loosdrecht M et al (2022) From wastewater to resource. One Earth 5(2):122–125. https://doi.org/10.1016/j.oneear.2022.01.011

Pruss-Ustun A, Wolf J, Bartram J, Clasen T, Cumming O, Freeman M et al (2019) Burden of disease from inadequate water, sanitation and hygiene for selected adverse health outcomes: an updated analysis with a focus on low- and middle-income countries. Int J Hyg Environ Health 222(5):765–777. https://doi.org/10.1016/j.ijheh.2019.05.004

Qadir M, Drechsel P, Jiménez B, Kim Y, Pramanik A, Mehta P et al (2020) Global and regional potential of wastewater as a water, nutrient and energy source. Nat Resour Forum 44:40–51. https://doi.org/10.1111/1477-8947.12187

Root R (2022) Feces and forests: why poor WASH is a threat to the environment. Devex WASH Works. https://www.devex.com/news/feces-and-forestswhy-poor-wash-is-a-threat-to-the-environment-102558. Accessed 31 Oct 2024

Ross I, Cumming C, Dreibelbis R, Adriano Z, Nala R, Greco G (2021) How does sanitation influence people's quality of life? Qualitative research in low-income areas of Maputo, Mozambique. Soc Sci Med 271:113709. https://doi.org/10.1016/j.socscimed.2021.113709

Saha N, Mollah M, Alam M, Rahman M (2016) Seasonal investigation of heavy metals in marine fishes captured from the Bay of Bengal and the implications for human health risk assessment. Food Control 70:110–118. https://doi.org/10.1016/j.foodcont.2016.05.040

Sato T, Qadir M, Yamamoto S, Endo T, Zahoor A (2013) Global, regional, and country level need for data on wastewater generation, treatment, and use. Agric Water Manage 130:1–13. https://doi.org/10.1016/j.agwat.2013.08.007

Singh B, Chakraborty A, Sehgal R (2023) A systematic review of industrial wastewater management: evaluating challenges and enablers. J Environ Manage 348:119230. https://doi.org/10.1016/j.jenvman.2023.119230

Srivastav L (2020) Chemical fertilizers and pesticides: role in groundwater contamination. In: Prasad M (ed) Agrochemical detection, treatment and remediation. Butterworth Heinemann, pp 143–159. https://doi.org/10.1016/B978-0-08-103017-2.00006-4

Thebo A, Drechsel P, Lambin E (2014) Global assessment of urban and peri-urban agriculture: irrigated and rainfed croplands. Environ Res 9:114002. https://doi.org/10.1088/1748-9326/9/11/114002

Thomas S, Gambrill M, Gilsdorf R, Diagne N (2018) 3 hard truths about the global sanitation crisis. World Bank Blogs. https://blogs.worldbank.org/water/3-hard-truths-about-global-sanitation-crisis. Accessed 31 Oct 2024

Trianni A, Negri M, Cagno E (2021) What factors affect the selection of industrial wastewater treatment configuration? J Environ Manage 285:112099. https://doi.org/10.1016/j.jenvman.2021.112099

Tuholske C, Halpern B, Blasco G, Villasenor J, Frazier M, Caylor K (2021) Mapping global inputs and impacts from of human sewage in coastal ecosystems. PLoS ONE 16(11):e0258898. https://doi.org/10.1371/journal.pone.0258898

UN-Habitat (2023) Global report on sanitation and wastewater management in cities and human settlements. Nairobi, Kenya

UN-Human Settlements Program, UN-Habitat and World Health Organization, WHO (2021) Progress on wastewater treatment–global status and acceleration needs for SDG indicator 6.3.1. Geneva, Switzerland

UNEP (2023) Wastewater—turning problem to solution. A UNEP rapid response assessment. Nairobi, Kenya. https://doi.org/10.59117/20.500.11822/43142

UNESCO (2020) The United Nations World Water Development Report 2020: water and climate change. https://www.unwater.org/publications/un-world-water-development-report-2020. Accessed 31 Oct 2024

UN-Water (2018) Water quality and wastewater. https://www.unwater.org/sites/default/files/app/uploads/2018/10/WaterFacts_water_and_watewater_sep2018.pdf. Accessed 24 Oct 2024

UN-Water (2021) The United Nations world water development report 2021: valuing water. UNESCO, Paris, France

UN-Water (2022) Progress on wastewater treatment (SDG target 6.3). https://www.sdg6data.org/en/indicator/6.3.1. Accessed 24 Oct 2024

Van Minh H, Nguyen-Viet H (2011) Economic aspects of sanitation in developing countries. Environ Health Insights 5:63–70. https://doi.org/10.4137/EHI.S8199

Ward M, Jones R, Brender J, De Kok T, Weyer P, Nolan B et al (2018) Drinking water nitrate and human health: an updated review. Int J Environ Res Public Health 15(7):1557. https://doi.org/10.3390/ijerph15071557

WaterAid (2021) Economic report: unlock trillions of dollars with clean water, decent toilets and hygiene. London, UK

Wear S, Thurber R (2015) Sewage pollution: mitigation is key for coral reef stewardship. Ann N Y Acad Sci 1355(1):15–30. https://doi.org/10.1111/nyas.12785

Wear S, Acuña V, McDonald R, Font C (2021) Sewage pollution, declining ecosystem health, and cross-sector collaboration. Biol Conserv 255:109010. https://doi.org/10.1016/j.biocon.2021.109010

WHO (2018) Guidelines on sanitation and health. https://apps.who.int/iris/bitstream/handle/10665/274939/9789241514705-eng.pdf. Accessed 31 Oct 2024

WHO and UNICEF (2021) Progress on household drinking water, sanitation and hygiene 2000–2020: five years into the SDGs. WHO/UNICEF Joint Monitoring Programme for Water Supply, Sanitation and Hygiene (JMP). https://www.who.int/publications/i/item/9789240030848. Accessed 31 Oct 2024

World Bank Group (2020) Wastewater a resource that can pay dividends for people, the environment and economies. https://www.worldbank.org/en/news/press-release/2020/03/19/wastewater-a-resource-that-can-pay-dividends-for-people-the-environment-and-economies-says-world-bank. Accessed 24 Oct 2024

World Bank, ILO, WaterAid, and WHO (2019) The health, safety and dignity of sanitation workers—an initial assessment. https://documents1.worldbank.org/curated/en/316451573511660715/pdf/Health-Safety-and-Dignity-of-Sanitation-Workers-An-Initial-Assessment.pdf. Accessed 31 Oct 2024

World Waterfall Database (2017) Niagara Falls, Ontario, Canada. https://www.worldwaterfalldatabase.com/index.php/waterfall/Niagara-Falls-106. Accessed 24 Oct 2024

Xu A, Wu Y, Chen Z, Wu G, Wu Q, Ling F et al (2020) Towards the new era of wastewater treatment of China: development history, current status, and future directions. Water Cycle 1:80–87. https://doi.org/10.1016/j.watcyc.2020.06.004

Xu X, Yang H, Li C (2022) Theoretical model and actual characteristics of air pollution affecting health cost: a review. Ijerph 19:3532. https://doi.org/10.3390/ijerph19063532

Zhongming Z, Linong L, Wangqiang Z, Wei L (2020) Sanitation and wastewater Atlas of Africa. https://www.unep.org/resources/publication/sanitation-and-wastewater-atlas-Africa. Accessed 24 Oct 2024

Chapter 2
Wastewater Generation and Management Patterns in Sub-Saharan Africa

Abstract In this chapter, wastewater (WW) generation and management plans in Sub-Saharan Africa (SSA) region were discussed. Generation rates in most countries were estimated at 0–10 m^3 yr^{-1} while collection, treatment and reuse of WW was below 5% in most countries with exception of South Africa and Nigeria in some cases. Low generation rates were associated with limited monitoring on WW production in the region due to the lack of sewer connections and infrastructure and hence, underestimation using available data. The low rates of collection, treatment and reuse were attributable to the lack of financial, technological and human resources to manage the effluents and incorporate it in a circular economy. In the evaluated case studies, WW management was characterized by the overflow and discharge of raw sewage, low priority input to manage the resource by specific governments, privatization of WW and sanitation services and extensive pollution, public health and environmental concerns from the mismanagement of the resource. The need to manage WW sustainably was emphasized to reverse its associated negative effects.

2.1 Introduction

In most African countries, the access to clean and safe water is essential since it influences environmental health and wellbeing of living things (Okesanya et al. 2024). In the continent just like the rest of the world, water scarcity and pollution are some of the most pressing issues confronting the residents and derailing the capacity of the region to realize sustainable development. Factors such as exponential population growth, industrialization, urbanization and inadequate treatment of wastewater (WW) have exacerbated the apparent trend (Nyika and Dinka 2023; Onu et al. 2023; Omohwovo 2024). Although advances and technologies have been developed to treat WW, Africa still struggles to manage the resource making it a growing environmental problem (Onu et al. 2023).

Unscientific management of WW has led to its non-treatment, accumulation in the environment and its unwanted release to natural resources (Torrens et al. 2020). The World Water Assessment Program (WWAP 2017) reported that about 90% of the

WW produced globally is discharged to the environment without adequate treatment. In another study, Kumari et al. (2023) noted that only 10% of the generated WW was processed and treated prior to discharge while the UN-Habitat and WHO (2021) reported that 40% of such sewage underwent treatment before its environmental release. Ravina et al. (2021) noted that high-income countries treat more than 70% of the generated WW while upper-middle income, lower-middle income and low-income countries only treat 38, 28 and 8% of their generated effluents, respectively. In Africa, only 1% of the generated WW is treated (Massoud et al. 2009). The rest of the WW is released untreated due to poor financing, institutional, technical and infrastructural capacity to manage the resource. Many African countries are culprits of discharging raw WW to the environment indiscriminately, which has grown to be a significant cause of environmental pollution and degradation. Most of the WW often results from domestic activities, industrial processes, from agricultural runoff and urban storm water runoff (Wang et al. 2014; Omohwovo 2024).

Efforts to manage natural resources including water resources and WW in Africa compared to population growth and urbanization trends are on a declining trend. Evidently, Africa's freshwater resources are at risk of degradation in the near future unless WW is well managed (Onu et al. 2023; Nyika and Dinka 2023). The impacts of water quality degradation will further extend to other socioeconomic sectors due to the healthcare burden from a high prevalence of waterborne and vector-bone diseases (Omohwovo 2024). Ultimately, the environment and public health sectors will be adversely affected (Ravina et al. 2021).

With the apparent effects associated with mismanagement of WW, there is need to prioritize on managing the effluents and their associated challenges in Africa (Nansubuga et al. 2016). This is particularly so in the Sub-Saharan Africa (SSA) region, which is vulnerable to water resources pollution (Nyika and Dinka 2023). Such advances include changing the current linear model of WW management to a circular model that prioritizes on recovery, reuse and maximal efficiency on WW recycling towards sustainability (Wang et al. 2014; Omohwovo 2024). Such a shift will enable the realization of sustainable development goals (SDGs). As noted by Ravina et al. (2021), non-prioritizing on WW management will compromise efforts to realize SDGs. This is particularly so for SDG 6.3 whose focus is on improving water quality through scientific management of WW and solid waste and ensuring their safe return to the environment.

Managing the shift begins with a clear understanding of the generation, management and treatment patterns of WW. Globally and more pronounced in SSA region, WW monitoring is inconsistent, infrequently done and the data on its production and management is scattered, unreported and unavailable (Mateo-Sagasta et al. 2015; Qadir et al. 2020; Nyika 2022). These challenges often hinder proper management of WW, its valorization and eventual recovery. This chapter therefore aims to explore on the generation and management patterns of WW in SSA region to provide clues on best practices to incorporate the resource in a circular economic model.

2.2 Overview of Sub-Saharan Africa Region

The SSA describes the region of Africa south of the Sahara desert, comprising of 48 countries and a population of approximately one billion people and a noticeably high population growth rate compared to the rest of the world (World Bank 2017). The region population is growing exponentially especially the urban population which is expected to double by 2050 (Brandoni and Bosnjakovic 2017). The region is endowed with many natural resources that have the capacity to push the realization of SDGs though most of the countries are low-income nations characterized by majority of the population dwelling in informal settlements where access to basic services including electricity, sanitation and clean waste is limited or even non-existent (Pariente 2017; Kanyerere et al. 2018). The growth patterns of the population accompanied by limited basic services is a setback to environmental sustainability and socio-economic growth in the region (Onu et al. 2023). The observation is attributable to emergent challenges in dealing with the high levels of solid waste and WW generation patterns.

Generation of WW and its unscientific discharge to the environment without treatment is more pronounced in SSA despite the region being one of the driest in the world (Darko et al. 2020). This tendency will promote water pollution and water scarcity. The situation is likely to worsen since WW management is not commensurate to population growth and urbanization patterns. Therefore, WW management should be prioritized since it will stimulate the realization of SDGs whose main driver is the availability of clean, adequate and safe water.

SSA region is lagging behind in attaining SDGs due to a myriad of problems including inadequate WW remediation, poor policies, institutions and political will to address the WW problem and inadequate water supply leading to water shortage (Niasse and Varis 2020). The WW problem is more pronounced in cities of SSA nations such as Ouagadougou, Niamey, Lagos, Kampala and Bamako that are capitals of Burkina Faso, Niger, Nigeria, Uganda and Mali, respectively (Chukwuma 2018). To this end, SSA region and leadership of individual nations must review their water situation with greater commitment from regulatory agencies and the government. Such a move would require emulating other dry countries such as Australia, North America, Saudi Arabia and Israel, which despite being water-stressed are able to collect and treat WW and use it as a resource for agricultural irrigation and in recharging of aquifers (Marin et al. 2017). Additionally, SSA region should commit to improve their water and WW infrastructure to eliminate combined sewer systems in preference to on-site treatment, increase funding to manage WW, enhance policy and administrative tools that support WW treatment, reclamation and reuse to valorize the resource (Maila et al. 2018).

2.3 Wastewater in Sub-Saharan Africa Region

Cognizant of the WW management problems and impacts of its discharge prior to treatment, many countries including those of SSA region are putting efforts to manage the resource and prevent its associated pollution. Concurrently, the efforts are promoting a circular economy, augmenting water supply sources and contributing to sustainable development positively (Nyika 2022). There are variations in the per capita generation of WW, its collection, treatment and recovery in many regions even among SSA countries based on the economic development (UNU-INWEH 2024).

Figure 2.1 shows the amount of WW generated in SSA in $m^3 \ yr^{-1}$ per capita in the SSA region while Fig. 2.2 shows the generation rates in million m^3 annually. Nations of the regions such as South Africa and Nigeria that are well off economically had higher production rates ranging from 50 to 100 $m^3 \ yr^{-1}$ per capita compared to other poorer nations. Jones et al. (2021) who reported that wealth and WW generation were positively correlated irrespective of the geographical location of focus confirmed the trend. In poor nations of SSA, the institutional, technical and infrastructural capacity to treat water and WW is limited due to financial challenges (Ravina et al. 2021) and hence the observed trend. Production rates as shown in Fig. 2.2 were mainly in urban rather than rural set ups of the region. The trend is associated with the discrepancies in providing sanitation services including sewerage collection and treatment in rural and urban SSA where the former is highly disfavored (Ali and Gujiba 2024).

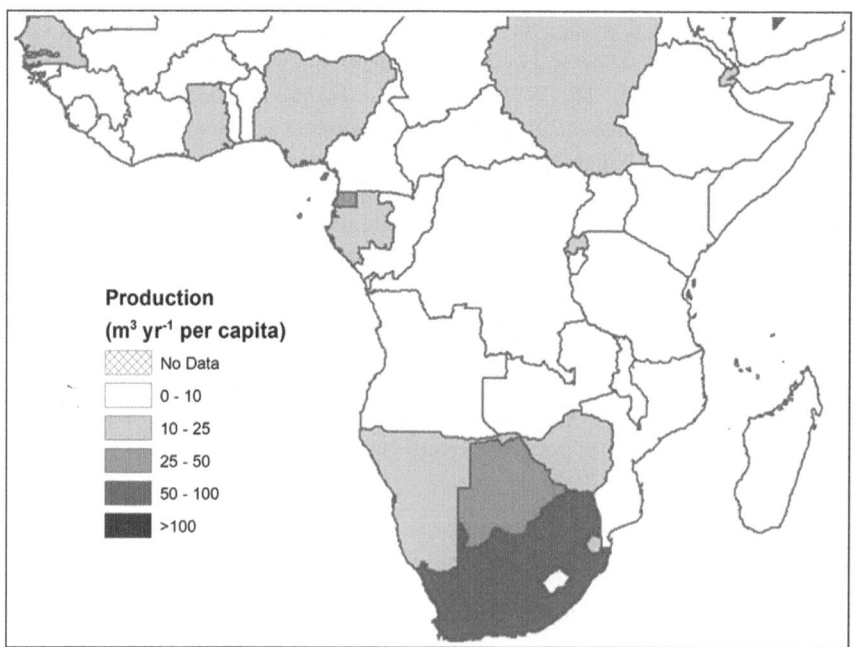

Fig. 2.1 Per capita wastewater production rates in SSA region in $m^3 \ yr^{-1}$ (UNU-INWEH 2024)

2.3 Wastewater in Sub-Saharan Africa Region

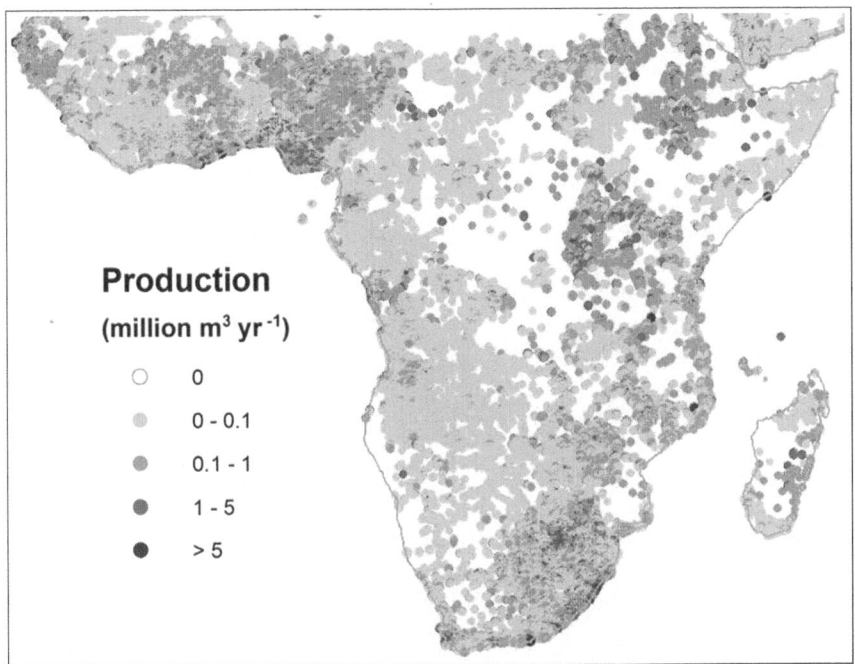

Fig. 2.2 Wastewater production rates in SSA region in million m³ per year (UNU-INWEH 2024)

Many of the countries reported low production rates of approximately $10\,m^3\,yr^{-1}$ per capita. The trend does not necessary depict the actual picture on WW production rates but rather an inadequacy in the collection, monitoring and information exchange of data on the resource. Several authors have reported that data on WW generation patterns in SSA is unreliable, incomplete, incomprehensive, inconsistent, infrequently reported and sometimes unavailable to make realistic inferences due to the lack of formal sewerage and sanitation systems and infrastructure in the region (Onu et al. 2023; Ali and Gujiba 2024; Drechsel et al. 2024). With the challenges, it is likely that the reported production rates have been underestimated.

Collection of generated WW in the region were evidently low (Fig. 2.3) as most countries reported a 0–5% collection rate (UNU-INWEH 2024). The apparent trend is attributable to a lacking infrastructure to collect, transport and treat both solid waste and WW in addition to poor planning in the sector, limited finances, lack of technical expertise and a negative public attitude to taking up localized onsite WW treatment systems (AfDB, UNEP and GRID-Arendal 2020). A similar trend was established by the UN-Habitat (2023), where SSA had the lowest collection rates compared to the rest of the world due to the use of conventional water supply systems. Urban South Africa and Nigeria had higher collection rates (Fig. 2.4) due to the development of WW management infrastructure associated with better economic endowment unlike other SSA nations (Jones et al. 2021).

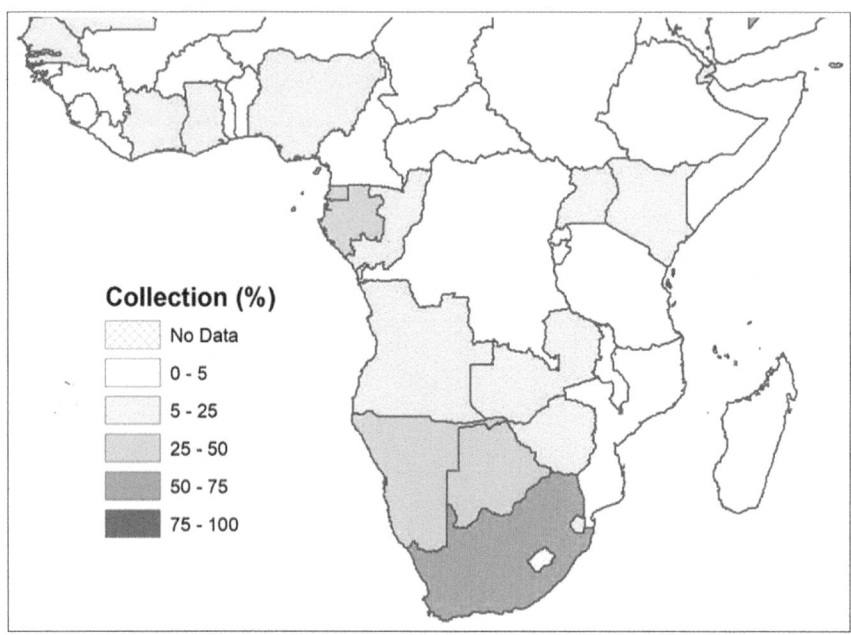

Fig. 2.3 Percentage wastewater collection rates in SSA (UNU-INWEH 2024)

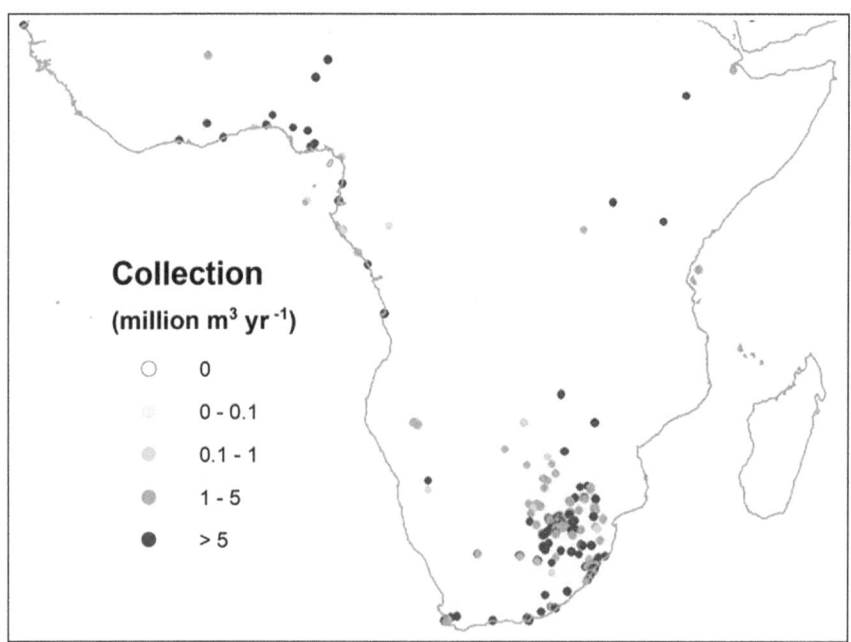

Fig. 2.4 Amount of wastewater collected in SSA region in million m^3 (UNU-INWEH 2024)

2.3 Wastewater in Sub-Saharan Africa Region

The treatment of WW in the region was low with most countries recording 0–5% rates with exception of a few such as South Africa that has at treatment range of 25–50% (Fig. 2.5). Similarly, South Africa and Nigeria had recorded a treatment capacity of between 1 and 5 million m^3 yearly (Fig. 2.6). Low rates of WW treatment is attributable to limited technical capacity whereby most plants in SSA use either conventional activated sludge or waste stabilization ponds as treatment techniques although they are less effective in remediating WW pollutants and require extensive land use that is also become scarce recently in the region (Rugaimukamu et al. 2022). Consequently, SSA countries including Senegal, Morocco, Burkina Faso and Ghana deal with increased WW flows that have high organic loads despite having low capacity to treat it due to high energy costs, poor WW recovery and limited plants operation and maintenance capacity (AfDB, UNEP and GRID-Arendal 2020).

Data available on tracking efforts to realize SDG target 6.3 on safe treatment of WW particularly from domestic sources shows that concerted efforts are needed to improve the practice. Between 2020 and 2023, SSA region only managed to treat 20% of its generated WW with some specific countries such as Benin and Central African Republic having only 1% of the resource channeled for treatment (UN-Water 2024). The reported proportion is lower than previously reported (24%) by Dinka and Nyika (2024) for the period between 2015 and 2020. The proportion of domestic WW flow that underwent safe treatment between 2020 and 2023 in specific African countries is as shown in Table 2.1 (UN-Water 2024). Evidently, SSA is off track

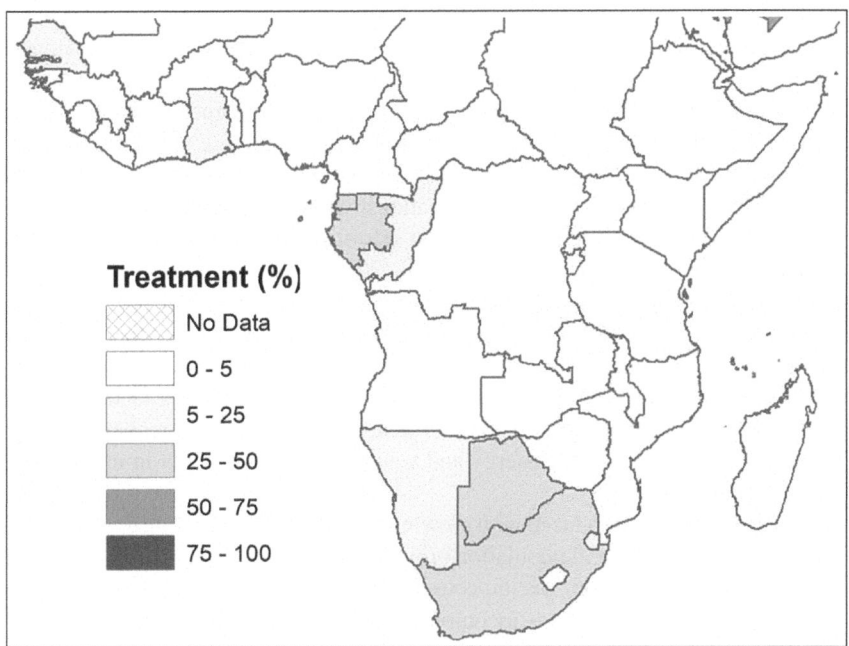

Fig. 2.5 Percentage wastewater treatment in SSA region (UNU-INWEH 2024)

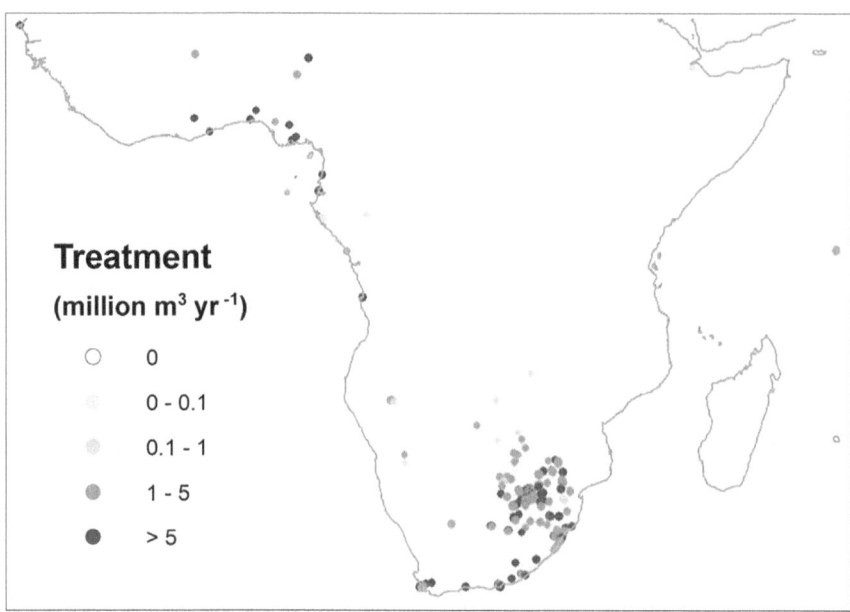

Fig. 2.6 Amount of treated wastewater in SSA region in million m^3 per year (UNU-INWEH 2024)

in realizing SDG 6.3 and therefore, improving the technical, financial and human resource capacity to treat municipal WW and industrial effluents (whose treatment is hardly reported) should be a prioritized (UN-Habitat 2023). Dinka and Nyika (2024) also suggested the need to expand connection to sewer lines, septic tanks and sanitation facilities to majority of the population to enhance safe effluent handling.

Reuse of WW in SSA region was low with exception of southern Africa (Figs. 2.7 and 2.8). Reuse was nearly 0% in many nations, which corresponded to the use of conventional water supplies and the limited infrastructure of WW in the region as reported by Qadir et al. (2020), Jones et al. (2021), and UN-Habitat (2023) who established similar patterns. Although pointing out data gaps and uncertainty in computing WW reuse, the authors point out that uptake of the practice is low in SSA despite its potential to relieve the region off water stress. WW reuse efforts should therefore be prioritized as has been done in the Arab countries such as United Arab Emirates and Jordan where 100% of generated effluents are rechanneled for reuse and recovery to enhance energy and water security especially in urban areas (Rugaimukamu et al. 2022).

Evidently, SSA region is faced with growing rates of WW generation due to urbanization, industrialization and population growth among other drivers. However, data on the generation patterns in specific countries remains sparse and where available, it has uncertainties that make it incomplete, unreliable and incomprehensible for making inferences (Onu et al. 2023). This is especially so in most countries, whose majority of the urban population dwell in informal settlements whereby water supply

2.3 Wastewater in Sub-Saharan Africa Region

Table 2.1 Proportion of safely management wastewater in named SSA nations based on data availability (UN-Water 2024)

Country	% of safely treated WW	Country	% of safely treated WW
Benin	1	Malawi	6
Burkina Faso	3	Mali	6
Central African Republic	1	Mauritius	19
Chad	2	Niger	7
Cote d'Ivoire	17	Nigeria	48
Democratic Republic of the Congo	16	Reunion	74
Djibouti	11	Senegal	14
Eswatini	17	Sierra Leone	15
Ethiopia	3	South Africa	61
Gambia	11	Togo	15
Ghana	12	Uganda	4
Guinea Bissau	21	Tanzania	8
Kenya	11	Zimbabwe	55
Madagascar	9	SSA	20

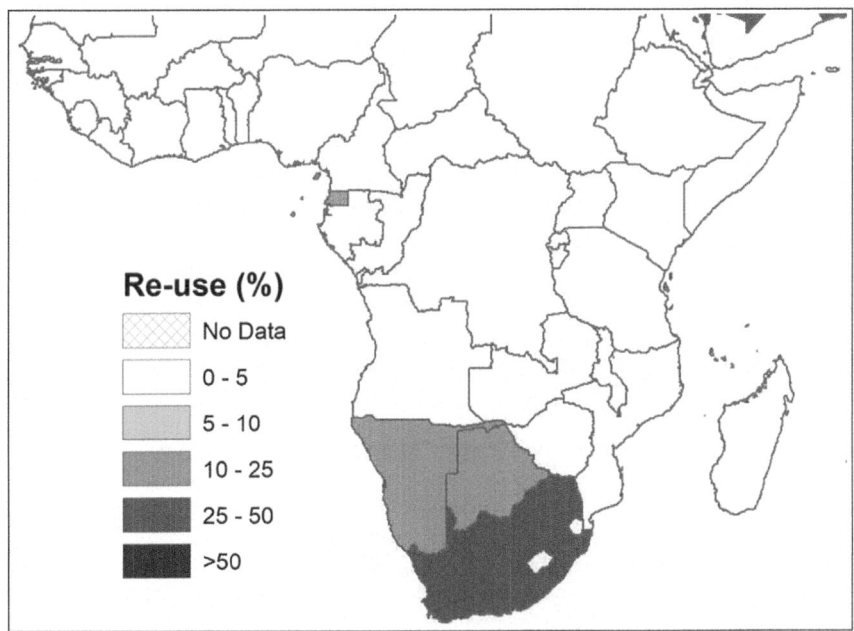

Fig. 2.7 Percentage wastewater reuse in SSA region (UNU-INWEH 2024)

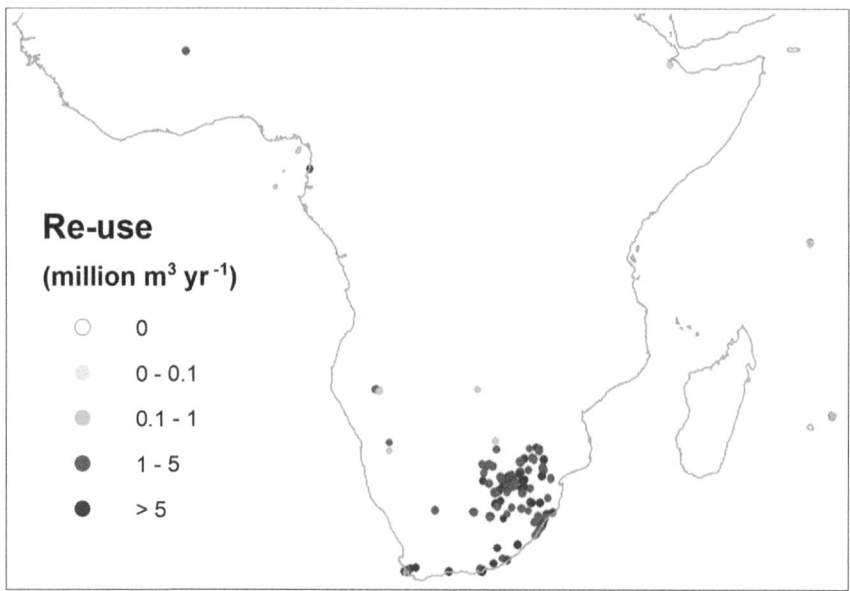

Fig. 2.8 Wastewater reuse rates in SSA in million m^3 per year (UNU-INWEH 2024)

is conventional, WW infrastructure is inadequate, and as such, no scientific risk management strategies are available to deal with the impacts of resource discharge prior to treatment (Drechsel et al. 2024). For these reasons, the region must shift to view WW as a resource by enhancing its circularity through proper collection, treatment and reuse. Encouraging WW reuse will improve environmental and human health, increase economic resilience and enhance recovery of the resource as well as encourage climate change mitigation and adaptation (Smol et al. 2020; Delgado et al. 2021). However, such initiatives must be supported by strong policy, institutional, and individual behavioral adjustments as well as political commitment to better the water, WW and solid waste sectors at all governance levels.

2.4 Case Studies of Wastewater Management in SSA Region

To understand the common aspects in water generation and management patterns in SSA region, this section reports on trends in specific countries. The countries of focus included Benin, Ethiopia, Malawi, Kenya and South Africa. The selection of the countries was based on availability of data on WW in specific countries.

2.4.1 Benin

Benin is a country in West Africa with a population of more than 14 million people (World Meter 2024a, b, c). Only 12.8% of its population has access to safe WW treatment services while majority of the populace have access to community latrines or have constructed their sewerage systems (Nation Institute of Statistic and Economic Analysis, INSAE 2016). Most of the generated municipal WW is disposed off to streets and courtyards ending up in the environment without any form of treatment (Ravina et al. 2021). As such, 70% of households release waste in nature while 20% dispose it in courtyards (INSAE 2016). Industrial effluents from urban areas are let out untreated in sewer lines and end up polluting the Atlantic Ocean (Nyambe et al. 2024).

Sanitation services in the country began being offered in 2003 and hence, the infrastructure is relatively young. These developments came as a result of the decentralization of WW services to municipalities so that sewer lines were constructed in towns and rural areas were equipped with toilets at schools and households for sanitation purposes (Daouda et al. 2021). According to Atinkpahoun et al. (2020), 61% of the urban population and 17% of the rural population have access to sanitation.

In the country's most populous and economic city, Cotonou, 57% of the households discharge WW in nature while 9.6, 7.2 and 3.5% dispose it in courtyards, closed and open sewers, respectively prior to any form of treatment (INSAE 2016). About 21% of the residents in the city have access to individual sanitation systems including sumps, latrines and collection tanks. Desludging of septic tanks and latrines was reported to be done mechanically in 94% of these facilities with 73% of the households doing it once a year (Hounkpe et al. 2014). Soak-away pits were also managed through manual desludging and 84% of greywater is discharged to the environment without any treatment (Hounkpe et al. 2014). Stabilization pond plants manage WW but are owned by private companies though effluent treatment is below standards.

With the management approaches, human health and environmental preservation is only done partially in the city, which is at high risk to pollution due to the infiltration of WW in the shallow water table and eventually to groundwater systems (Atinkpahoun et al. 2020). The trend is also replicated countrywide as UN-Water (2024) noted that only 1% of all the generated water is safely treated. With inadequate management, environmental discharge of WW has polluted freshwater systems including ground- and surface-water introducing disease causing pathogens and multi-resistant germs, which result to waterborne infections (Onifade et al. 2017; Adanlokonon et al. 2018). The privatization of sanitation and WW management has further complicated efforts to treat the resource as such private companies are largely unregulated and unmonitored.

2.4.2 Ethiopia

Ethiopia is a landlocked country in East Africa with a population of more than 133 million people (World Meter 2024a, b, c). The country has seen growth in urbanization as its population increases and hence, a growth in WW generation. Although this is the prevailing trend just like other SSA countries, WW treatment facilities in the nation are nearly non-existent and the ones present, are poorly managed (Haddis et al. 2013). Even in large cities such as Addis Ababa, poor drainage characterized by overflow of effluents from institutions, industries and residential areas are predominant. According to Ebissa et al. (2024), the city produces more than 49 million m^3 of WW daily out of which, 13.4% is from industrial effluent and only 14,000 m^3 undergoes treatment out of the total generated. Teferi (2014) also reported that only 45% of the population in Addis Ababa in 2014 had access to safe water supply mainly due to its contamination by WW.

The discharge of untreated WW to natural waterways and streams in Ethiopia can be directly correlated to limited sanitation. In the country, 73% of urban and 77% of rural population uses unimproved sanitation facilities (WHO/UNICEF Joint Monitoring Program 2021). Open defecation was reported to be practiced by 43% of rural and 8% of urban population. Although in larger cities, regional and state capitals, sanitation facilities are partially covered, such privileges are not there in other rural parts of the country (Ravina et al. 2021). The priority in many urban areas is to enhance access to improved toilets but the collection and treatment of WW prior to discharge is still low. More than 60% of urban households use traditional pit latrines and more than 5% are using open defecation (Ravina et al. 2021). Managing WW is conventional whereby fecal sludge is accumulated in poorly built and designed pits and then it is discharged into open water bodies, storm drains or allowed to infiltrate into aquifer systems. In other instances, the fecal sludge is removed from pit latrines manually for discharge to the environment.

Even more recent statistics show that the country only has 3% of its WW being treated safely (UN-Water 2024), which is a public health and environmental concern. Consequently, freshwater systems in the country have low water quality and this has exacerbated the water stress situation in the already dry Ethiopia (Worku 2018). Adoption of conventional WW treatment technologies such as biological treatment and up-flow anaerobic sludge blanket especially in Kaliti wastewater treatment plant in Addis Ababa among other cities has resulted to extended environment pollution from resultant toxic sludge leading to the damage of biodiversity and pollution of agricultural soils (Shuralla et al. 2024). A potential human health risk is imminent following the use of untreated WW for irrigating vegetables especially in urban Ethiopia and in the form of waterborne diseases from drinking freshwater contaminated with WW (Geshaye 2020). Evidently, the country needs to invest more finances, technology and human resources to ensure WW generation is monitored countrywide and its collection, treatment and reuse is done effectively and safely. Such advances require the collaborations of involved regulatory agencies, ministries, local governments, the private sector and individual WW producers.

2.4.3 Malawi

Malawi is one of the poorest and the most densely populated nations of SSA with a population of more than 21 million people. It is located in the south, has a population density of 129 people per km^2 and an urban growth rate of 6.3% yearly (Ngoma et al. 2020; Ravina et al. 2021). Approximately 15% of the population uses septic tanks to dispose WW from industrial and domestic sources while another 15% has access to waterborne sewage (Msilimba and Wanda 2021). Approximately 93.2, 1.4 and 5.4% use pit latrines, econ-san toilets and flush toilets connected to septic tanks and sewer lines, respectively (Holm et al. 2021). Urban areas are well deserved on WW management compared to rural areas. Cities such as Mzuzu, Lilongwe, Zomba and Blantyre have centralized treatment plants whereby 34% of the 66% liquid WW is safely managed (Ravina et al. 2021). In more recent statistics of 2020–2023, only 6% of all the WW generated in the country is safely treated (UN-Water 2024). Collet et al. (2018, 2021) reported that only 1% of the 10% WW collected is treated off-site while 8% is no treated at all in the Blantyre city. In Kasungu town, only 1% of 5% collected WW is treated off-site while the rest is not delivered or treated at the plant (Collet et al. 2018, 2021).

Like many SSA nations, majority of Malawians discharge untreated WW into natural waterways and storm drains while a minority discharge the resource to septic tanks installed in their households or in bare soil for collection by municipalities (Ravina et al. 2021). According to the UN-Habitat and WHO (2021), Malawi produces 211.9 million m^3 of total household WW whereby out of the total 5.5, 8.8 and 85.6% ends up in sewers, septic tanks and other forms of sanitation, respectively. Of the total generated only 21 million m^3 is collected and 13.7 million m^3 is treated safely, which represents only 6.5% of total WW (UN-Habitat and WHO 2021). Once pit latrines are full especially in rural areas, they are abandoned and others are constructed which can be a hazard if the fecal matter infiltrates in areas with shallow groundwater. Private companies and city councils also operate business of collecting and emptying WW from pit latrines and septic tanks to channel them to sewer lines in urban areas. The handling of the WW amidst rising pressures from growing industrialization, urbanization and population growth is overwhelming the municipalities and available private collectors.

With only a small proportion of the generated WW being collected and treated, the risk of environmental pollution, waterborne diseases and biodiversity loss is high in Malawi. This is also evident in other SSA countries and hence the need to change the WW management approaches (Ravina et al. 2021; Ngoma et al. 2020). Such changes incorporate frequent effluent monitoring, improved WW infrastructure and strong policies and institutions to reform the sector and infuse it in the circular economy.

2.4.4 Kenya

Kenya is an Eastern African country with a population of more than 56 million people. The country's urban and suburban population is growing and so is the demand for clean water and sanitation services (Angatia 2013). Consequently, WW generation from industries, toilets, bathrooms and kitchens is increasing. According to a report by UN-Habitat and WHO (2021) the country produces about 832 million m^3 of household WW and out of the figure 12.8, 11.9 and 75.3% is collected in sewers, septic tanks and other sanitation means, respectively. However, due to unavailability and paucity of data, actual proportion of WW that is collected, treated and reused is not available (Angatia 2013). Most of the generated WW is discharged on roadside channels, open fields, open ponds and septic tanks just as in Ethiopia and the rest of SSA. Using the approaches, the WW accumulates foul smells, becomes a breeding ground for mosquitoes among other pathogens that result to malaria and waterborne diseases. Urban residents are partially covered in accessing WW services unlike rural areas. In Kenya's capital, the generated WW is collected from septic tanks and pit latrines by private collectors and/or the Nairobi county government and taken to decentralized treatment plants at Ruai and Kariobangi for conventional treatment (Kilingo et al. 2021). Other approaches to manage WW include using stabilization ponds, aerated lagoons and constructed wetlands.

In addition to the risk of diseases, the mismanagement of WW characterized by inadequate service provision and limited infrastructural development on the WW sector has induced widespread pollution of ground- and surface-water resources and deteriorated the environmental health in general (Angatia 2013; Kilingo et al. 2021). Crops and soils irrigated with such WW also take in some of the pollutants in the resource resulting to their extensive spread in food chains (Nyika and Dinka 2022). Informal settlers dwelling in urban areas especially in marginal and low-lying land along polluted drainage channels are severely affected by WW mismanagement.

2.4.5 South Africa

South Africa has a population of more than 60 million people and is the largest economy in SSA region (Mustapha 2025). Compared to the rest of the globe, the country ranks 123rd out the possible 187 in the human development index that includes SDG 6 on clean water and sanitation (Water Research Commission, WRC 2021). About 41% of its population is not connected to sewer systems and more than 20 million people in the country use unsafe sanitation facilities including bucket, chemical and pit latrines while more than 2 million people have no access to any formal sanitation facilities (WRC 2021). Consequently, production of WW is on a rising trend and as such, more than 50% of rivers and 40% of surface water bodies lack ambient water quality due to discharge of raw sewage from industrial, municipal and agricultural sources (WRC 2021).

In the country, the awareness of the need to manage WW sustainably has also grown along with economic development. Consequently, much attention and financial investment is being committed to improve WW treatment technology and in engineering of treatment plants. According to Edokpayi et al. (2020), South Africa spends more than 3 billion rand per year on treatment of the resource. Aoyi et al. (2015) reported that the country has about 950 WW processing plants. Approaches such as aeration basins, trickling filters, anaerobic ponds, stabilization ponds and enhanced biological nutrient removal (ENBR) have been applied to manage effluents (Edokpayi et al. 2020).

However, even with high spending on treatment infrastructure, quality compliance of treated effluents has been a challenge (Edokpayi et al. 2020). Failures in WW management in the country just like other SSA nations are associated with inadequate maintenance of existing infrastructure, limited human resources, underserved rural residents and poor planning for the rapidly expanding urban settings (WRC 2021). More than 67% of treatment facilities across the country do not comply with the predefined regulatory standards despite majority having a low treatment capacity of 500–2000 m^3/day, which is an environmental and public health risk (Department of Water Affairs, DWA 2012). Archer et al. (2017) made similar observations highlighting that the inadequate treatment and subsequent release of effluents to the environment in South Africa has resulted to hypertrophic rivers and surface waters polluted with emerging contaminants. The tendencies have exacerbated the apparent water scarcity state in the country.

Uptake of treated WW reuse is documented and practiced in the country because of the role the resource plays in relieving the pressure of freshwater over-use and reducing high nutrient content in surface water bodies (Edokpayi et al. 2020). The water is reused for urban and peri-urban agriculture and in some cases, as potable water (Jaramillo and Restrepo 2017). WW reuse is a result of concerted efforts by local authorities, communities and specific industry initiatives. A case example is the collaboration between Emalahleni municipality, the local community and Anglo coal, BHP Billiton company to treat mine water and supplement domestic water supply by making the WW potable (Edokpayi et al. 2020).

To overcome challenges in WW management, the country must alleviate the discharge of raw effluents. Such efforts can be realized through better coverage of WW services, monitoring for compliance in treating the resource, improvement of aging infrastructure and better financial and human resources input in the sector (Edokpayi et al. 2020; WRC 2021). The situation in South Africa is replicable in most SSA countries that suffer from similar WW management challenges.

2.5 Discussion and Conclusion

The presented data on wastewater generation, collection, treatment and reuse in SSA region and the five country-specific case studies discussed confirmed that sanitation is challenging and still lacking. Although most countries reported lower WW generation levels (0–0.1 million m^3 yr^{-1}), the data was scattered and possibly underestimated due to limited infrastructure and the use of conventional management approaches. Holistically, SSA region only treated 20% of its WW safely while collection and reuse was low (0–5%) in most nations with exception of South Africa. Most of the generated WW was discharged to the environment untreated and below the stipulated standards.

The unscientific discharge of WW exacerbated by the lack of sanitation facilities presented in most countries pose a public health and environmental threat. In urban areas where WW treatment plants, (mainly centralized systems) are present, they are not properly managed and most of the effluent is eventually discharged without meeting the predefined standards. In addition, the treatment capacity of the plants is inadequate, their technology application and efficiency monitoring is lacking compared to apparently high WW generation rates. Ultimately, efforts to realize SDGs are negatively affected.

To reverse the dire state in SSA region, this chapter highlights the need to enhance resource treatment capacity by applying better technology and infrastructure in managing the WW cycle. Enhancement of fecal sludge management through the use of advanced sanitation facilities to preserve human and environmental health should be prioritized. Monitoring of WW production, collection, treatment and reuse patterns should be revamped in all SSA countries to enable data collection, continuous monitoring of operations and applied technology, decision making on improvements and affirmative action in lacking areas. Such advances require more human, financial and technological resources' investment.

References

Adanlokonon E, Kanhounnon W, Chabi B, Adjahouinou D, Koumolou L, Bonou B et al (2018) Physicochemical and microbiological characterization of effluents from the "Centre Hospitalier Universitaire de la mere et del'E nfant lagune (CHU-MEL) discharged in the Cotonou lagoon in Benin. Int J Biol Chem Sci 12(4):1955–1964

African Development Bank, United Nations Environment Program and GRID-Arendal (2020) Sanitation and wastewater atlas of Africa. AfDB, UNEP and GRID-Arendal, Abidjan, Nairobi and Arendal

Ali A, Gujiba U (2024) Household wastewater management in sub-Saharan Africa: a review. Discov Water 4:6. https://doi.org/10.1007/s43832-024-00060-6

Angatia P (2013) Factors influencing wastewater management and re-use in peri-urban areas in Kenya: a case study of Ongata Rongai. MA Thesis, University of Nairobi, Kenya

Aoyi O, Onyango M, Majozi T, Seid E, Eswifi T, Rwanga S, Kesi J (2015) Water and wastewater management in local government: skills needs and development final report part II to the Local Government Sector Education and Training (LGSETA)

Archer E, Gideon M, van Wyk J (2017) Pharmaceutical and personal care products (PPCPs) as endocrine disrupting contaminants (EDCs) in South African surface waters. Water SA 43(4):684–706. https://doi.org/10.4314/wsa.v43i4.16

Atinkpahoun C, Pons M, Louis P, Leclerc J, Soclo H (2020) Rare earth elements (REE) in the urban wastewater of Cotonou (Benin, West Africa). Chemosphere 251:126398. https://doi.org/10.1016/j.chemosphere.2020.126398

References

Brandoni C, Bosnjakovic B (2017) Homer analysis of the water and renewable energy nexus for water-stressed urban areas in Sub-Saharan Africa. J Clean Prod 155:105–118. https://doi.org/10.1016/j.jclepro.2016.07.114

Chukwuma O (2018) Rural water supply in Nigeria: policy gaps and future directions. Water Policy 20(3):597–616. https://doi.org/10.2166/wp.2018.129

Collet S, Tilley E, Yesaya M (2018) SFD report; Centre for Water, Sanitation, Health and Appropriate Technology Development (WASHTED), Zomba, Malawi

Collet S, Tilley E, Yesaya M (2021) SFD report. https://www.susana.org/_resources/documents/default/3-3545-7-1550665329.pdf. Accessed 6 Nov 2024

Daouda M, Hounkpe P, Djihouessi B, Akowanou O, Aina M, Drogui P (2021) Physicochemical assessment of urban wastewater of Cotonou (Benin). Water Sci Technol 83(6):1499–1510. https://doi.org/10.2166/wst.2021.073

Darko R, Liu J, Yuan S, Sam-Amoah L, Yan H (2020) Irrigated agriculture for food self-sufficiency in the sub-Saharan African region. Int J Agric Biol Eng 13(3):1–12. https://doi.org/10.25165/j.ijabe.20201303.4397

Delgado A, Rodriguez D, Amadei C, Makino M (2021) Water in circular economy and resilience. World Bank, Washington, DC. http://hdl.handle.net/10986/36254

Dinka M, Nyika J (2024) SDG 6 progress analyses in sub-Saharan Africa from 2015–2020: the need for urgent action. Discov Water 4:39. https://doi.org/10.1007/s43832-024-00099-5

Drechsel P, Bartram J, Qadir M, Medlicott K (2024) The challenge of supporting and monitoring safe wastewater use in agriculture in LMIC. NPJ Clean Water 6(7):1–3. https://doi.org/10.1038/s41545-024-00364-z

DWA (Department of Water Affairs, South Africa) (2012) Green drop progress report 2012. http://www.dwaf.gov.za/dir_ws/GDS/Docs/DocsDefault.asp. Accessed 8 Jan 2025

Ebissa G, Fetene A, Desta H (2024) Study on quality of treated wastewater for urban agriculture use in Addis Ababa, Ethiopia. City Environ Interact 24:100157. https://doi.org/10.1016/j.cacint.2024.100157

Edokpayi J, Folami A, Adeeyo A, Durowoju O, Jegede A, Odiyo J (2020) Recent trends and national policies for water provision and wastewater treatment in South Africa. In: Singh P, Milshina Y, Tian K, Gusain D, Bassin J (eds) Water conservation and wastewater treatment in BRICS nations. Elsevier. https://doi.org/10.1016/B978-0-12-818339-7.00009-6

Geshaye D (2020) Wastewater-irrigated urban vegetable farming in Ethiopia: a review on their potential contamination and health effects. Cogent Food Agric 6:1772629. https://doi.org/10.1080/23311932.2020.1772629

Haddis A, Geyter A, Smets I, Bruggen B (2013) Wastewater management in Ethiopian higher learning institutions: functionality, sustainability and policy context. J Environ Plan Manage 57(3):369–383. https://doi.org/10.1080/09640568.2012.745396

Holm R, Chunga B, Mallory A, Hutchings P, Parker A (2021) A qualitative study of NIMBYism for waste in smaller urban areas of a low-income country, Mzuzu, Malawi. Environ Health Insights 15:117863022098414. https://doi.org/10.1177/1178630220984147

Hounkpe S, Adjovi E, Crapper M, Crapper M, Awuah E (2014) Wastewater management in third world cities: case study of Cotonou, Benin. J Environ Prot 5(5):387–399. https://doi.org/10.4236/jep.2014.55042

INSAE (2016) Effectifs de la population des villages et quartiers de ville du Benin (RGPH-4, 2013)

Jaramillo M, Restrepo I (2017) Wastewater reuse in agriculture: a review about its limitations and benefits. Sustainability 9(10):1734. https://doi.org/10.3390/su9101734

Jones E, Van Vliet M, Qadir M, Bierkens M (2021) Country-level and gridded estimates of wastewater production, collection, treatment and reuse. Earth Syst Sci Data 13(2):237–254. https://doi.org/10.5194/essd-13-237-2021

Kanyerere T, Tramberend S, Levine A, Mokoena P, Mensah P, Chingombe W et al (2018) Water futures and solutions: options to enhance water security in sub-Saharan Africa. In: Systems analysis approach for complex global challenges. Springer, pp 93–111. https://doi.org/10.1007/978-3-319-71486-8_6

Kilingo F, Zulu B, Chen H (2021) The analysis of wastewater treatment system efficiencies in Kenya: a review paper. Int J Sci Res Publ 11(5):204–215. https://doi.org/10.29322/IJSRP.11.05.2021.p11322

Kumari S, Dwivedi S, Khan M, Nayanam S, Dhasmana A, Malik S (2023) The challenges of wastewater and wastewater management. In: Shah M (eds) Advanced and innovative approaches of environmental biotechnology in industrial wastewater treatment. Springer, Singapore. https://doi.org/10.1007/978-981-99-2598-8_5

Maila D, Mathebula V, Crafford J, Mulders J, Eatwell K (2018) Towards the development of economic policy instruments for sustainable management of water resources. Water Research Commission, Pretoria, South Africa

Marin P, Tal S, Yeres J, Ringskog K (2017) Water management in Israel: key innovations and lessons learned for water-scarce countries. World Bank Group, Washington DC

Massoud M, Tarhini A, Nasr J (2009) Decentralized approaches to wastewater treatment and management: applicability in developing countries. J Environ Manage 90(1):652–659. https://doi.org/10.1016/j.jenvman.2008.07.001

Mateo-Sagasta J, Raschid-Sally L, Thebo A (2015) Global wastewater and sludge production, treatment and use. In: Drechsel P, Qadir M, Wichelns D (eds) Wastewater: economic asset in an urbanizing world. Springer Netherlands, Dordrecht, pp 15–38. https://doi.org/10.1007/978-94-017-9545-6_2

Msilimba G, Wanda E (2021) Wastewater production, treatment, and use in Malawi. https://www.ais.unwater.org/ais/pluginfile.php/231/mod_page/content/188/country_report_malawi.pdf. Accessed 6 Nov 2024

Mustapha J (2025) Sub-Saharan Africa. Published online at futures.issafrica.org. https://futures.issafrica.org/geographic/regions/sub-saharan-africa/. Accessed 8 Jan 2025

Nansubuga I, Banadda N, Verstraete W, Rabaey K (2016) A review of sustainable sanitation systems in Africa. Rev Environ Sci Biotechnol 15:465–478. https://doi.org/10.1007/s11157-016-9400-3

Ngoma W, Hoko Z, Misi S, Chidya R (2020) Assessment of efficiency of a decentralized wastewater treatment plant at Mzuzu University, Mzuzu, Malawi. Phys Chem Earth (Pt A, B, C) 118–119:102903. https://doi.org/10.1016/j.pce.2020.102903

Niasse M, Varis O (2020) Quenching the thirst of rapidly growing and water-insecure cities in sub-Saharan Africa. Int J Water Resour Dev 36(2–3):505–527. https://doi.org/10.1080/07900627.2019.1707073

Nyambe N, Ngigi M, Nderu J, Awoyodo C (2024) Review of innovative approaches of wastewater treatment and management: a case study of Benin, Kenya and Zambia in pursuit of sustainable solutions in the context of the green deal. J Civ Eng Res Technol 6(3):1–11. https://doi.org/10.47363/jcert/2024(6)159

Nyika J (2022) Wastewater for agricultural production, benefits, risks, and limitations. In: Chatoui H, Merzouki M, Moummou H, Tilaoui M, Saadaoui N, Brhich A (eds) Nutrition and human health. Springer, Cham. https://doi.org/10.1007/978-3-030-93971-7_6

Nyika J, Dinka M (2022) Heavy metal pollution in soils and vegetables from suburban regions of Nairobi Kenya and their community health implications. Pollution 8(4):1434–1447. https://doi.org/10.22059/poll.2022.341522.1440

Nyika J, Dinka M (2023) Water challenges in rural and urban Sub-Saharan Africa and their management. Springer Nature, Cham, Switzerland

Okesanya O, Eshun G, Ukoaka M, Manirambona E, Olabode O, Adesola R et al (2024) Water, sanitation, and hygiene (WASH) practices in Africa: exploring the effects on public health and sustainable development plans. Trop Med Health 52:68. https://doi.org/10.1186/s41182-024-00614-3

Omohwovo E (2024) Wastewater management in Africa: challenges and recommendations. Environ Health Insights 18:1–6. https://doi.org/10.1177/11786302241289681

References

Onifade S, Baba-moussa F, Aïna M, Noumavo P, Toukourou F (2017) Physico-chemical and bacteriological characterization of surface waters (well, pond and lagoon) and industrial wastewater in Cotonou City (Benin). Int J Sci Environ Technol 6(2):1001–1018

Onu M, Ayeleru O, Oboirien B, Olubambi P (2023) Challenges of wastewater generation and management in sub-Saharan Africa: a review. Environ Chall 11:100686. https://doi.org/10.1016/j.envc.2023.100686

Pariente W (2017) Urbanization in sub-Saharan Africa and the challenge of access to basic services. J Demogr Econ 83(1):31

Qadir M, Drechsel P, Jiménez B, Kim Y, Pramanik A, Mehta P et al (2020) Global and regional potential of wastewater as a water, nutrient and energy source. Nat Resour Forum 44:40–51. https://doi.org/10.1111/1477-8947.12187

Ravina M, Galletta S, Dagbetin A, Kamaleldin O, Mng'ombe M, Mnyenyembe L et al (2021) Urban wastewater treatment in African countries: evidence from the hydro aid initiative. Sustainability 13:12828. https://doi.org/10.3390/su132212828

Rugaimukamu Q, Wang H, Huang R, Xie L (2022) Wastewater treatment needs more attention. Nat Afr. https://doi.org/10.1038/d44148-022-00071-2

Shuralla A, Hiruey A, Gebreeyessus G (2024) Treatment appraisal and fate of HMs in up-flow anaerobic sludge blanket and trickling filter-based sewage treatment process: the case of a Kaliti centralized wastewater treatment plant, Addis Ababa, Ethiopia. Heliyon 10(13):e34003. https://doi.org/10.1016/j.heliyon.2024.e34003

Smol M, Adam C, Preisner M (2020) Circular economy model framework in the European water and wastewater sector. J Mater Cycles Waste Manage 22(3):682–697. https://doi.org/10.1007/s10163-019-00960-z

Teferi Z (2014) Written communication on water management of Addis Ababa

Torrens A, Varga D, Ndiaye A, Folch M, Coly A (2020) Innovative multistage constructed wetland for municipal wastewater treatment and reuse for agriculture in Senegal. Water 12(11):3139. https://doi.org/10.3390/w12113139

UN-Habitat (2023) Global report on sanitation and wastewater management in cities and human settlements. Nairobi, Kenya

UN Habitat and WHO (2021) Progress on wastewater treatment—global status and acceleration needs for SDG indicator 6.3.1. United Nations Human Settlements Program (UN-Habitat) and World Health Organization (WHO), Geneva

United Nations University-Institute for Water, Environment and Health (UNU-INWEH) (2024) Global wastewater status. Retrieved from: Global Wastewater Status | United Nations University. Accessed 5 Nov 2024

UN-Water (2024) Sub-Saharan Africa. Retrieved from: Region | SDG 6 Data. Accessed 5 Nov 2024

Wang H, Wang T, Zhang B, Li F, Toure B, Omosa I, Chiramba T et al (2014) Water and wastewater treatment in Africa—current practices and challenges. Clean 42:1029–1035. https://doi.org/10.1002/clen.201300208

Water Research Commission, WRC (2021) The status of wastewater as an untapped resource in South Africa. Pretoria, South Africa

Worku H (2018) Rethinking urban water management in Addis Ababa in the face of climate change: an urgent need to transform traditional to sustainable system. Environ Qual Manage 27(1):103–119. https://doi.org/10.1002/tqem.21512

World Bank (2017) Atlas of sustainable development goals 2017: from world development indicators. The World Bank

World Health Organization (WHO) and The United Nations Children's Fund (UNICEF) (2021) Progress on household drinking water, sanitation and hygiene 2000–2020: five years into the SDGs. UNICEF, New York, NY, USA

World Meter (2024a) Benin. https://www.worldometers.info/world-population/benin-population/. Accessed 5 Nov 2024

World Meter (2024b) Ethiopia population. https://www.worldometers.info/world-population/ethiopia-population/. Accessed 6 Nov 2024

World Meter (2024c) Kenya population. https://www.worldometers.info/world-population/kenya-population/. Accessed 6 Nov 2024

WWAP (2017) Wastewater: the untapped resource, the United Nations World Water Development Report. The United Nations World Water Assessment Program

Chapter 3
Wastewater Treatment and Management in SSA

Abstract This chapter explored on the treatment technologies applied to manage wastewater (WW) in Sub-Saharan Africa (SSA). Both conventional and advanced technologies of WW treatment were applied in specific countries of the region. The efficacy of conventional approaches to remove nutrient loads from WW was high though for other pollutants, it was low and effluents required further treatment. The techniques including septic systems and waste stabilization ponds required large space for operation and were prone to producing sludge. Greener approaches such as the use of biodigesters and composting toilets were effective in managing WW especially Blackwater and converted it to useful energy and bio-fertilizer sources. However, their implementation in SSA was at small-scale due to their high cost and maintenance needs. The need to optimize these treatment techniques to treat more WW was emphasized as generation rates increased in the region and concerns on effluent management and its negative implications were growing despite the apparent low treatment capacity of the region. Such initiatives need to be supported with the right financial, technological, human and infrastructural resources.

3.1 Introduction

Clean water is a fundamental component for a healthy ecosystem, abundant biodiversity, enhanced public health and human wellbeing in Africa (Bateganya et al. 2015). Being the second driest continent globally after Australia, the continent's total renewable water resources are 9% and support more than 15% of the world's population (Bedair et al. 2023). The water scarcity situation of the region is further aggravated by expanding urbanization, industrialization and population growth tendencies, which put pressure on sanitation facilities and the ability to manage freshwater and wastewater (WW) (Bateganya et al. 2015; Nwokediegwu et al. 2024). Additionally, the tendencies promote unplanned and uncontrolled settlements and set up of industries, which enhance the use of freshwater and production of large quantities of WW, that is discharged to the environment indiscriminately leading to extensive pollution (Dinka and Nyika 2024).

Current efforts to manage the increasing levels of WW in Africa are not commensurate with expansion of urban areas, the growth of industries and population growth (UN-Habitat 2023). Therefore, most of the WW in the continent is discharged and reused without any form of treatment leading to environmental pollution especially to downstream surface waters (Wang et al. 2014; Kjellen 2018; Onu et al. 2023). Among rural and urban population, WW is essential for food security and shapes many livelihoods but this is not without human health and environmental risks. According to Mishra et al. (2023), WW is used for nourishing crops, car washing, firefighting, in building construction, watering golf courses and as a coolant in power plants especially in arid and water stressed nations of Africa and across the globe. Such uses are imperative in meeting sustainable development goals (SDGs) including those related to poverty reduction, food security enhancement, better health and wellbeing, access to clean water and sanitation and energy security among others (Obaideen et al. 2023; Silva 2023). However, maximal benefits of WW uses can be realized if effective treatment is done.

In Africa and particularly, Sub-Saharan Africa (SSA), WW collection and treatment is limited and in most cases not documented (UN-Habitat 2023). This is despite the crucial role of the resource's treatment in enhancing public health, environmental sustainability and effective water resources management. SSA suffers from regulatory framework limitations, financial constraints and infrastructural gaps that hinder the deployment of appropriate technology to collect and treat WW leading to its discharge (Nwokediegwu et al. 2024). The release and eventual reuse of WW prior to effective treatment results to environmental, health and economic impacts that are counterproductive to efforts to realize SDGs. Such impacts include modifying the physicochemical characteristics of soils and their productive capacity, polluting freshwater resources, exposing living things including humans to pathogens and heavy metal poisoning and their associated complications (Mishra et al. 2023).

To counter the negative effects of using WW in a region experiencing a growing demand for clean water such as SSA, it is imperative to manage the resource by understanding how it can be treated and reused safely (Onu et al. 2023). Appropriate technology application for pollutant remediation from WW and regulations to ensure safe discharge and reuse of resource also need to be elucidated and applied in SSA as has been done in Australia and North Africa (Gomez-Sanabria et al. 2020; Miarov et al. 2020). In the countries, the generated WW is recycled and reused for various consumptive uses. However, in SSA, WW management monitoring is hardly done and available data on the resource is incomplete, incomprehensive, unreliable and in some cases, non-existent (Ali and Gujiba 2024; Drechsel et al. 2024). To bridge the existent gaps, this chapter explores on the treatment methods applied in managing WW and the treatment capacity in SSA using specific countries as examples.

3.2 Wastewater Infrastructure in SSA

In SSA, improving the access to clean, good quality and safe drinking water and sanitation is a great challenge in the region due to lack of standardized infrastructure network to supply water, collect and treat WW (Onu et al. 2023). In the latest statistics of 2022, only 31% of the population in SSA uses drinking water that is safely managed while 24% of the population has access to safely managed sanitation services (UN-Water 2024). Lack of access to safe drinking water is exacerbated by freshwater resources pollution and the introduction of WW in the resources due to limited use of safely managed sanitation facilities. A further 80% of all the generated domestic WW in SSA is not safely treated and as such, discharged without any form of treatment (UN-Water 2024).

The apparent trends are attributable to poor water infrastructure in the region: only, 6, 15 and 14% use surface water, unimproved, and limited drinking water services (UN-Water 2024). In this case surface water is directly fetched from irrigation channels, canals, streams, dams, lakes and rivers while unimproved sources are unprotected springs and wells, which in most cases are unsafe (WHO 2024). Limited drinking water sources are from an improved source whereby water seekers take more than 30 min to complete a roundtrip of fetching water. Lack of safely managed drinking water sources in SSA is attributable to ageing assets and inadequate infrastructure to supply water especially in urban areas (UNEP 2023; George-Williams et al. 2024). As such, it is difficult to access, store, regulate, circulate and conserve water resources in the region due to these challenges. Additionally, separating freshwater and WW networks is also impossible with the infrastructural limitations (Dangui and Jia 2022). As such, most of the used water is contaminated with WW.

About 17% of the SSA population practice open defection while 31 and 17% use unimproved and limited sanitation facilities (UN-Water 2024). Unimproved sanitation facilities refers to pit latrines that do not have a platform or slab, bucket latrines and hanging latrines while limited facilities refer to improved sanitation facilities that are shared between and among households. Improved sanitation facilities are hygienic and separate black water from human contact. In Nigeria alone more than 47 million people were reported to be practicing open defecation, which has severe repercussions on the environment, natural resources and public health (Mara 2017; Onifade 2024). Kenya (Muriuki et al. 2020) and Ethiopia (Ebissa et al. 2024) were also reported to have poorly designed WW systems whose sewer line connectivity were poor and as such, most effluents were released in open drains.

The limitation in provision of sanitation is because of the lack of sewer networks. A study by Lerebours et al. (2021) in 20 SSA cities reported that over 80% of the populace in the countries uses onsite sanitation facilities whereby toilets and latrines have no direct connection to sewer line and onsite containment occurs followed by emptying of the fecal sludge and its transfer to a centralized WW treatment plant. Furthermore, the transportation of the WW with fecal sludge from the source to the treatment plant is unsafe characterized by leaks that lead to extensive environmental pollution (Lerebours et al. 2021). Atangana and Oberholster (2023) reported similar tendencies in a study assessing the progress of sanitation in SSA as the region strives to realize SDGs. In Uganda, dilapidated and degraded infrastructure was attributable to the leaks in municipal sewers of Masaka municipality and the lack of WW generation and treatment monitoring and data collection (Bateganya et al. 2015).

Apparent discrepancies in access to improved water and WW infrastructure exists in rural and urban SSA where the former is undeserved compared to the latter. According the UN-Water (2024), in 2020, 53% of urbanites in SSA had access to safely managed drinking water while only 15% of the rural population accessed such services. Additionally, only 30% of urbanites accessed safely managed sanitation services while 20% rural residents had access to such services in the same period. The discrepancies are a result of superior infrastructure development in urban areas compared to rural areas. According to Atangana and Oberholster (2023), SSA rural areas and urban slums lack secure and adequate housing with improve water, solid waste and sewerage collection and hence the apparent trends. Moreover, most rural SSA is cut off from accessing safe water and WW networks and access only polluted drinking water and at the same time, discharge untreated WW in freshwater systems (Omohwovo 2024). Ravina et al. (2021) established a similar trend while looking at the status and coverage of WW services in SSA countries of Malawi, Ethiopia and Benin. In Nigeria, urban WW infrastructure is obsolete and not designed for the current population. Additionally, it is not updated or maintained frequently while in most rural areas; such basic infrastructure is not there (Onifade 2024).

The lack of formal sanitation and sewerage services in most of SSA is justified by the economic situation of the region. As one the poorest regions of the world, SSA is not able to keep up with the capital-intensive nature of WW infrastructure worsened by pressures of increased generation due to drivers such as urbanization and population growth. According to Onu et al. (2023), water and WW infrastructure is capital demanding initially which is followed by a prolonged payback period that SSA countries cannot afford. A report by the UNEP (2023) also observed that poor countries such as those in the SSA region could not finance the expensive WW infrastructure and technology since they have limited public funds to transit from centralized and low-tech management approaches to decentralized and tech-savvy systems of managing sewage used in developed countries. SSA also lacks private participation in infrastructure (PPI), which is an essential driver in funding water and WW infrastructure especially in developed countries (World Bank 2018, 2019). Without PPI, up-to-date infrastructure and sanitation facilities cannot be set up leading to low productivity of water and WW projects and ultimately, the lack of sustainable development (Onu et al. 2023).

In a country like Ghana, WW infrastructure is barely allocated any resources and therefore infrastructure, human, financial and technological capacity to manage the resource is limited (Sackey et al. 2023). To reverse the situation, the governments and regulatory agencies in the region must cultivate the maintenance culture and intensify funding on water and WW infrastructure to prevent the negative effects of its mismanagement on their economies, environment and the human wellbeing at large (UNEP 2023). Additional efforts should be made to improve the technical and institutional capacity to handle the ever-increasing WW generation levels from not only domestic but also from industrial sources (Ravina et al. 2021; UNEP 2023).

3.3 Wastewater Treatment Techniques

Treatment of WW describes the processes and methods used to remove pollutants and contaminants from effluents in their aqueous form. In SSA, the handling of WW is either centralized or decentralized (Ali and Gujiba 2024). The former refers to collection and treatment of sewage for larger communities while the latter, refers to treatment of WW at individual buildings and/ or homes. Centralized systems require sewer networks to collect WW from households and industries for treatment at large-scale plants, which are often publicly owned and situated away from the generation points (Nansubuga et al. 2016). On the other hand, decentralized systems are localized and treatment and disposal of effluents is close to the generation points. The preference to either centralized or decentralized WW treatment systems is dependent on cost implications and the ecological sensitivity of the effluent generators. Decentralized WW treatment systems are cheap to maintain especially in remote areas with sparse population compared to centralized systems that are often located in densely populated urban settings (Nansubuga et al. 2016; Ali and Gujiba 2024).

The common treatment approaches are physical, chemical, biological and sludge remediation (Dutta et al. 2021; Musa and Idrus 2021). In physical treatment approaches, methods such as aeration, filtration, skimming, sedimentation and screening of WW occurs (Bhargava 2016). In chemical treatment, bacteria and other pathogens in WW are destroyed using chemicals such as ozone and chlorine, addition of bases and acids that stabilize the pH of effluents (Bhargava 2016; Crini and Lichtfouse 2019). Biological approaches add microorganisms to degrade organic contaminant that are in WW in composting, aerobic and anaerobic processes (Gupta et al. 2018).

Adsorbents including biosorbents such as activated charcoal and plant biomass waste also take up pollutants from WW due to their high adsorptive capacity (Islam et al. 2019). WW treatment via adsorption can be combined with nanotechnology to enhance absorption and adsorption of pollutants and the removal efficacy (Onu et al. 2023). Sludge treatment also referred to us as activated sludge processing involved the conversion of non-settable components of WW into settable ones (Tiwari and Awasthi 2022). Treating sludge reduces WW storage space, its associated transport cost and enhances treatment efficiency by employing several methods. The applied methods include thickening, dewatering, conditioning and drying for volume reduction, biological, chemical and electrochemical stabilization processes before the final byproduct is disposed (Yadav et al. 2022).

The method or combination of methods to treat WW depends on effluent characteristics and the desired use of the water after treatment (Steyn et al. 2021). In most conventional cases, WW management involves the successive steps of preliminary, primary, secondary and tertiary treatment before managing sludge (Crini and Lichtfouse 2019). In preliminary treatment, grit, large floating materials and oily substances are removed while in primary treatment, settleable inorganic and organic solids as well as suspended solids (SS) are removed via skimming and sedimentation. In secondary treatment, suspended biosolids are degraded using activated sludge and aeration techniques while tertiary treatment uses advanced technology applying chemicals to purify the water via approaches such as oxidation, ozonation, chlorination and coagulation among others. The degree of WW treatment influence the reuse of the resource. For pretreated and primary treated WW, reuse is not recommended due to safety concerns. However, secondary treated WW can be used as an industrial coolant, in recharge of groundwater and for surface irrigation while tertiary treated water can be used for cleaning vehicles, irrigation and in toilet flushing (Maryam and Buyukgungor 2019). Portable use of WW is only done if advanced technology is employed to treat the resource.

In SSA region, some of the WW treatment techniques are applied but at different scales. The technique depends on the desired treatment degree but also the complexity associated with WW pollutants some of which are refractory in nature (Edokpayi et al. 2017; Tetteh et al. 2019). Treatment approaches can either be on-site or off-site (Wang et al. 2014). On-site treatment approaches include the use of septic tanks and pit latrines, which in most cases are poorly maintained, end up overflowing and being public health and environmental hazards. Off-site WW treatment approaches are common in urban areas where collected WW is transported to centralized treatment plants for management. Some of the conventional and advanced approaches to WW treatment are discussed in the following sub-sections and illustrated in Figs. 3.1 and 3.2.

3.3.1 Septic Systems

Septic systems are the commonest approach to manage WW (Wang et al. 2014). They consist of an underground chamber or tank made of plastic, fiberglass and concrete whereby domestic WW flows for treatment via aerobic digestion and settling, which reduce organic content and solids in the sewage. Septic systems have been used to remove SS, grease, chemical and biochemical oxygen demand (COD and BOD). They can be installed in individual households and have low maintenance cost but their treatment efficiency is low and hence, the need to further treat WW via secondary treatment approaches (Onu et al. 2023).

In SSA, the proportion of household WW produced in septic tanks is 24.8% of all produced water and accounts for 16,259 million m^3 of all generated sewage (UN-Habitat and WHO 2021). The proportion of water that ends up in septic systems in specific SSA countries is as shown in Table 3.1. Angola, Guinea Bissau, Nigeria, Senegal and Togo were some of the SSA countries that had a high preference for using septic tanks with 64, 44.4, 49.8, 47.9 and 60.1%, respectively of their WW ending up

3.3 Wastewater Treatment Techniques

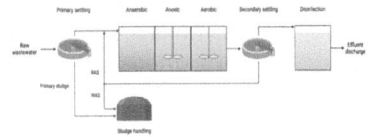

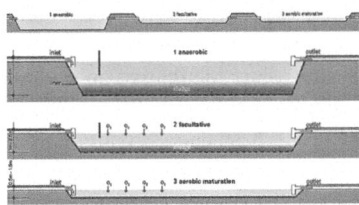

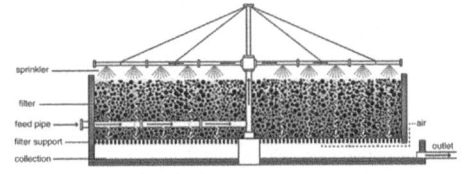

Fig. 3.1 Conventional approaches to WW treatment in SSA region. By Authors

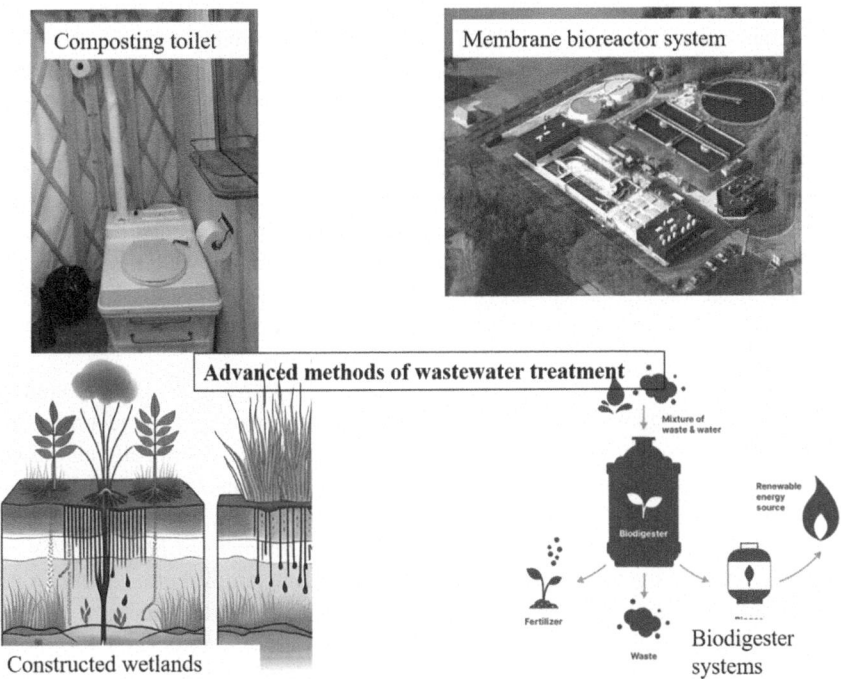

Fig. 3.2 Advanced methods of wastewater treatment in SSA region. Drawn by Authors

in such facilities. South Africa and Nigeria recorded the highest percentage of safely treated WW from septic systems at 91.8 and 75.1%, respectively. Other large WW generators such as the Democratic Republic of the Congo still had low rates of safely treated WW due to the aforementioned infrastructural limitations in the sector.

3.3.2 Waste Stabilization Ponds

Water Stabilization Ponds (WSP) are big shallow basins surrounded by embankments in which WW is treated biologically through natural processes using bacteria and algal species. They are designed as a series of anaerobic, facultative and maturation ponds where WW is flowing in and out continuously (Oberlin 2018). Anaerobic ponds process WW with high organic loads, are 2–5 m deep, maintained at a temperature of 15 °C and a pH of <6.2 and enable digestion of organic content reducing BOD levels in a period of 1–2 days (Oberlin 2018). In facultative ponds, BOD and SS are removed using natural growing algae via photosynthesis in a period of 5–30 days (Oberlin 2018). Maturation ponds enable further treatment of black water after passing through anaerobic and facultative ponds whereby they remove pathogens at high pH, solar radiation and high temperature (Oberlin 2018).

WSP are cost effective forms of WW treatment useful in many urban areas of SSA region due to their low power and maintenance cost but they are often too overloaded due to large quantities of sewage (Sinn and Lackner 2020; Sinn et al. 2022). In Rwanda, Tanzania, Nigeria, South Africa, Ghana, Malawi, Ethiopia and Kenya, WSP are common modes of physical and biological treatment of WW (Bansah and Suglo 2016; Oberlin 2018; Nwosu and Chukwueloka 2020). In Kenya's capital, Nairobi WSP are used to manage WW at the Dandora oxygenation ponds that handle more than 80,000 m^3/day of effluent (Wang et al. 2014). In Namibia, WSP modified with upstream anaerobic sludge blanket (UASB) pretreatment technology was applied to remediate COD, SS and *Escherichia coli* from WW (Sinn and Lackner 2020; Sinn et al. 2022).

The removal efficacy for SS, COD and BOD from WW using WSP was found to be 98.3, 92.8, and 92.4% in a study assessing the performance of the technique in pollutant remediation in Mwanza, Tanzania (Oberlin 2018). An overall efficiency rate of 77.78% to remove BOD, COD, SS, total coliforms, nutrients, (N and P), ammonia and chloride was reported using WSP in North Central Nigeria (Mi et al. 2022). In another study assessing the performance of WSP in treatment of hospital WW in rural South Africa, the technique was found to have a low removal efficacy for COD (<50%) and microbial parameters (<1 log reduction rates) (Edokpayi et al. 2021). The poor performance of the ponds in pollutant removal was associated with poor maintenance of the system due to lack of non-adherence to treatment guidelines and limited desludging. Overall, WSP require large land space and constant desludging (Onu et al. 2023) despite their efficiency in nutrient and organic load removal from WW.

3.3 Wastewater Treatment Techniques

Table 3.1 Proportion of water that ends up in septic systems in named SSA countries (UN-Habitat and WHO 2021)

Country	Proportion of WW generated (million m^3)	% of WW ending up in septic tanks	% of the WW from septic systems that is safely treated
Angola	566.8	64	–
Benin	158.1	11.9	–
Botswana	70.1	5.8	–
Burkina Faso	243.6	5.9	–
Burundi	106.3	16	–
Cameroon	423.6	28.5	–
Central African Republic	36.9	0.6	44.6
Chad	137.6	3.3	34.8
Comoros	21.7	7.7	–
Congo	117.7	24.6	–
Cote d'Ivoire	499.1	32.0	–
Democratic Republic of the Congo	11,019.6	28.1	41.8
Djibouti	19.2	20.6	30.5
Eritrea	55.9	11.6	–
Eswatini	23.9	12.4	44.2
Ethiopia	1356.1	6.9	–
Gabon	58.9	0.0	–
Gambia	45.9	41.8	23.1
Ghana	557.2	38.3	23.7
Guinea	238.3	23.8	–
Guinea Bissau	25.5	44.4	44.7
Kenya	831.8	11.9	–
Lesotho	30.7	2.9	–
Madagascar	348.9	17.2	44.7
Malawi	211.9	8.8	42.1
Mali	332.7	8.1	–
Mauritania	88.9	28.2	–
Mauritius	66.7	6.7	36.6
Mozambique	482.2	22.1	–
Namibia	60.9	3.1	–
Niger	264.3	16.4	18.2
Nigeria	2962.4	49.8	75.1
Sao Tome & Principe	3.6	15.2	–

(continued)

Table 3.1 (continued)

Country	Proportion of WW generated (million m³)	% of WW ending up in septic tanks	% of the WW from septic systems that is safely treated
Senegal	449.2	47.9	19.8
Sierra Leone	83.4	19.3	37.7
Somalia	261.4	9.1	–
South Africa	1700.1	3.4	91.8
Togo	95.6	60.1	24.5
Uganda	490.1	6.4	–
Tanzania	978.5	20.7	–
Zambia	269.4	18.2	-
Zimbabwe	115.9	6.0	49.2

3.3.3 Conventional Activated Sludge

Conventional activated sludge process (CASP) systems are used in biological treatment of WW. The system uses microbes that grow in sewage to form flocs or activated sludge that is suspended in the resource (Nancharaiah and Reddy 2018). CASP systems run continuously in sewage treatment facilities and operate in two distinct tanks. The first is an aeration tank where nitrification and organic carbon removal occurs and the second is a settling tank that separates treated water from activated sludge via flocculation (Nancharaiah and Reddy 2018). Further removal of nutrient is done biologically. The process is used to remove nutrients, SS, and BOD from agricultural and domestic effluents and requires less land space, is odor free and results to high pollutant remediation efficacy (Onu et al. 2023).

In SSA region, CASP, septic systems and WSP technologies are applied to manage over 70% of the total generated WW due to their low operational and maintenance costs (Rugaimukamu et al. 2022). In South Africa, CASP was used in secondary WW treatment in combination with sedimentation, rotating biological contractors and bio-filters (Okeyo et al. 2018). The technique was also used to manage effluents and sewage in Nigeria (Oloruntoba and Alabi 2019), Zimbabwe (Thebe and Mangore 2012), Uganda and Zambia (Wang et al. 2014) especially in urban areas. Although CASP systems are widely applied in SSA, high cost is incurred in managing the resultant sludge and its associated secondary pollution (Edokpayi et al. 2017). Additionally, monitoring of its efficiency in pollutant removal is hardly done.

3.3.4 Trickling Filters

Trickling filters (TF) are aerobic systems of managing WW that use microorganisms to remove organic matter. TF is part of attached growth processes such as packed bed reactors and rotating biological contractors whereby anaerobic, aerobic and facultative bacteria as well as protozoa, algae and fungi attached on a slim layer or biological film are used to treat WW (USEPA 2000). Removal of pollutants from WW occurs via sloughing where microbes on the layer or film attach and grow until they lose their ability to cling on the medium and then fall off for collection in an underdrain where they are transported for removal from WW (USEPA 2000). The process is simple and reliable in cleaning the WW off BOD and nutrients but further treatment is required to remediate refractory pollutants in sewage. TF have been used in WW treatment in Kenya, Zambia (Wang et al. 2014), South Africa (Wilsenach et al. 2013), and Zimbabwe (Thebe and Mangore 2012). Although the filters are simple to operate, they pose the risk of clogging depending on the characteristics of WW (Onu et al. 2023).

3.3.5 Constructed Wetlands

Constructed wetlands (CWs) are man-made engineered WW treatment systems designed to use microbial assembly, soils, wetland vegetation and natural processes to remediate pollutants through a combination of physicochemical and biological processes akin those of natural wetland ecosystems (Makopondo et al. 2020). In most cases, they are designed as shallow ponds within which aquatic plants are grown and once WW is passed, its contaminants are removed via either plant uptake or accelerated microbial and chemical degradation. CWs are ideal for water treatment in SSA region since they are cheap to design and set in addition to requiring minimum operational costs (Avellan and Gremillion 2019). The systems' set up also require less space compared to the conventional WW treatment approaches. Such wetland system use crops such as the *Cyperus papyrus, Typha spp., Phragmites spp., Lemma spp., and Eichhoria spp* (Avellan and Gremillion 2019). CWs are classified based on the macrophytes grown in them or based on the flow regime where they occur as surface or sub-surface systems. The three main kinds of CWs are subsurface, free water subsurface and hybrid systems and depending on the flow of water, each of the three kinds could be vertical or horizontal (Islam et al. 2022). The systems are useful in the removal of SS, chemical oxygen demand and nutrient content (P and N) from agricultural and domestic wastewater (Wang et al. 2014).

CWs were used in Kenyan resorts, eco-lodges and game lodges that were remotely located to manage WW and promote environmental conservation and sustainable tourism and hospitality operations (Makopondo et al. 2020). However, monitoring on their efficacy was challenging due to the lack of technical expertise and support to run the systems. Urban domestic greywater in Burkina Faso was managed using constructed wetlands that use local plant species to reduce fecal coliforms and remove SS, nutrients, COD and BOD from the sewage (Compaore et al. 2023).

The free water surface CWs system was employed in Nyanza, Kenya to manage sugar factory WW while subsurface flow systems were used to treat municipal wastewater in South Africa (Waly et al. 2022) Juja, Kenya, Dares Salam, Tanzania and Kampala, Uganda (Zhang et al. 2014)with varied success in removal of SS, nutrients and organic demand content of the sewage. In rural Burkina Faso, a subsurface horizontal flow CWs system was used to remove organic matter, orthophosphate and ammonia from greywater by more than 50% and reduce the level of fecal bacteria up to the WHO guidelines in order to reuse the resource for agricultural irrigation (Maiga et al. 2024). Other authors have documented the effective use of CWs in Kenya (Kilingo et al. 2021), Rwanda (Shyaka and Niyonzima 2019) and Senegal (Torrens-Armengol et al. 2020) due to its low maintenance and energy requirement though its pollutant removal efficacy is varied and not ideal.

3.3.6 Composting Toilets

To manage domestic WW containing black water, composting toilets are used to reduce pathogens in the sewage and enable the recycling of its nutrients by converting it to fertilizer (Onu et al. 2023). Composting toilets or dry toilets (they use little volumes or no water to flush) convert excreta and urine to organic matter by treating it at elevated temperatures to accelerate decomposition and thereafter, the resultant bio-fertilizer is used for agriculture (Odey et al. 2019). The conversion to bio-fertilizer occurs as a result of anaerobic decomposition. The toilets are suitable for areas deprived off sewerage connections, WW and waste treatment facilities as and alternative to flush toilets due to their low water requirement (Kubba 2012). In West African countries including Ghana (Odey et al. 2019; Mariwah et al. 2022), Ethiopia (Bidira et al. 2024) and other SSA countries (Armah et al. 2018), composting toilets are used not only as community-led total sanitation systems but also in the management of black water at household level.

3.3.7 Biodigesters

Biodigesters are anaerobic systems used in treating wastewater containing human excreta, agricultural and animal WW converting the wastes to biogas while cleansing sewage off SS, BOD and COD (Onu et al. 2023). According to Singh et al. (2022), anaerobic treatment can be done on sewage and effluents from municipal, industrial, and pharmaceutical sources and uses bioreactors such as anaerobic dynamic, anaerobic membrane, anaerobic filters, and anaerobic baffled reactors as well as expanded granular sludge bed and up-flow anaerobic sludge blanket. Apart from producing biogas, which is used as a fuel, biodigesters generate sludge.

In South Africa, the systems were used to manage sugarcane industry WW as reported by Mehrizi et al. (2023). Wastewater from Kenyan prisons, coffee factories, slaughterhouses and flower farms was treated using biodigesters as part of decentralizing effluent treatment (Energizing Development 2024). The onsite treatment contributes to energy security but such WW harnesses low levels of biogas compared to animal manure and as such, is less preferred (IEA 2015). In Benin cities of West Africa, WW management using biodigesters and its associated contribution to energy security has the potential to contribute to economic gains, improved sanitation and public health of the citizens (Mensah et al. 2021). Despite the fact that biodigesters encourage energy production and resource recycling, their operation and maintenance is complex, and they have low removal efficacy for nutrients (Onu et al. 2023).

3.3.8 Membrane Systems

Membranes are barriers that separate two phases through selective restriction of movement (Ezugbe and Rathilal 2020). Pressure driven membrane processes including nanofiltration (NF), ultrafiltration (UF), reverse osmosis (RO) and membrane filtration (MF), which are made of synthetic organic polymers have been applied to treat WW (Ezugbe and Rathilal 2020). The differences in the techniques are the membrane pore sizes and their pressure requirements. They are helpful in WW pretreatment but can also be combined with biological processes to remove organic and inorganic molecules in polluted water.

Ultrafiltration and MF techniques have been used in WW treatment in Namibia (Capodaglio 2021) and Nigeria (Oloruntoba and Alabi 2019). In South Africa, membrane bioreactor systems, use of RO and UF has revolutionized municipal and industrial WW treatment by being able to eliminate their pathogens and SS and ultimately, enabling the reuse of the effluents (Ion Exchange 2023; Othman et al. 2021). The methods are effective for WW treatment although they are cost, power, operations and maintenance intensive (Onu et al. 2023). Additionally, the techniques are prone to fouling and clogging and require constant cleaning and monitoring (Othman et al. 2021).

Evidently, WW treatment in SSA is mainly conventional characterized by pretreatment, primary, secondary and tertiary processes using techniques such as collection of sewage in septic tanks, use of stabilization ponds and via CASP (Rugaimukamu et al. 2022). Advanced WW techniques are also being used due to the growing awareness on the need to treat the resource though at small scales attributable to costs of the technologies. The treatment can occur both on and off site and its aim is to safely discharge WW and reduce its associated negative effects on the environmental and public health. Different techniques have different WW treatment capacities and pollutant removal efficacies.

3.4 Wastewater Treatment Capacity

The aim of WW treatment is to remediate its contaminants so that it is released to the environment safely for reuse (Kim et al. 2019). The techniques discussed in the previous sub-sections show that biological, physical, chemical and sludge activation processes are used in the treatment of WW. The capacity of treatment is a complex function of several factors including the specific technique of treatment, the equipment capacity, size of vessels, WW quantities generated, availability of treatment infrastructure and operation aspects (Gu et al. 2017). At country level, income is the key determining factor of treatment capacity such that high-income, lower-middle-income and lower-income countries have a 70, 30 and 10% WW treatment capacity, respectively (Drechsel et al. 2015). In another study, high-income, upper-middle income, lower-middle-income and lower-income countries had a 90, 70, 60 and 50% WW treatment capacity (Jones et al. 2021). The WWAP (2017) established that high-income, upper-middle income, lower-middle-income and lower-income countries had a 70, 38, 28 and 8% WW treatment capacity. Most of SSA countries are lower-income nations and hence, their low WW treatment capacity (Drechsel et al. 2015; Jones et al. 2021). The region has WW treatment capacity below 10% (Onu et al. 2023) and 5% (Akebe et al. 2023) despite the apparent drivers of population growth, industrialization and expansion of urban areas that put pressure on the already obsolete WW management infrastructure. Being low income countries, their WW treatment capacity is limited as such capability is positively correlated to income. Majority of the SSA countries have paid low attention to WW treatment and management since they consider it a non-priority issue despite their high capability and hence, the low treatment capacity (Onu et al. 2023; Ali and Gujiba 2024). The countries prioritize on buying weapons to oppress peoples' right other than improve their living conditions.

The region also has the lowest volumetric flow rates of collected, treated and recovered WW at <2 billion m^3 annually compared to regions such as East Asia Pacific and North America whose flow rate is approximately 12 and 9 billion m^3 annually (Jones et al. 2021). Figoli et al. (2014) noted that the region has WW treatment plants though their design accommodate low flows of WW ranging between 200 and 5000 m^3/h. Majority of the treatment plants for effluents are set up to comply with existent regulations other than to make WW safe before discharge and for environmental protection. As such, WW is treated at primary levels without prioritizing effective pollutant remediation (Blum et al. 2020). The financial, technical and human resources to improve the capacity of SSA WW treatment plants to manage the resource is limited (Onu et al. 2023). Therefore, upscaling WW infrastructure should be precedence to enable better WW management capacity.

3.5 Conclusion

This chapter found the application of both conventional and advanced WW treatment techniques in SSA. Some of the conventional approaches include WSP, CASP and septic systems useful in removal of organic pollutant load from effluents. The techniques are effective in cleansing WW off nutrients though they require a larger operation space and are not effective for refractory contaminants. The techniques also induce sludge production, which results to secondary pollution. Advanced technologies to manage SSA sewage were found to be CWs, composting toilets, membrane and biodigester systems. The techniques are greener and have a higher removal efficacy compared to conventional ones though they are costly, require frequent monitoring and are prone to fouling and clogging when treating WW. Depending on the effluent characteristics, the desired use of WW after treatment, cost and infrastructural implications, different techniques can be applied at different scales on-site and off-site. The treatment capacity of the region was found to be low at 5–10% of the total generated WW due to financial and infrastructural limitations of the member countries. The chapter recommended on the need to optimize the WW techniques to enhance their treatment efficacy and capacity without inducing secondary pollution through application of better technologies and improvement of effluent management infrastructure.

References

Akebe K, Afsatou T, Potgieter N (2023) Wastewater is a valuable source of information-Africa's scientists need to use it to find drug-resistant bacteria. The Conversation Africa, South Africa

Ali A, Gujiba U (2024) Household wastewater management in sub-Saharan Africa: a review. Discov Water 4:6. https://doi.org/10.1007/s43832-024-00060-6

Armah F, Ekumah B, Yawson D, Odoi J, Afitiri A, Nyieku F (2018) Access to Improved Water and Sanitation in Sub-Saharan Africa in a Quarter Century 4(11):e00931. https://doi.org/10.1016/j.heliyon.2018.e00931

Atangana E, Oberholster P (2023) Assessment of water, sanitation, and hygiene target and theoretical modeling to determine sanitation success in sub-Saharan Africa. Environ Dev Sustain 25:13353–13377. https://doi.org/10.1007/s10668-022-02620-z

Avellan T, Gremillion P (2019) Constructed wetlands for resource recovery in developing countries. Renew Sustain Energy Rev 99:42–57. https://doi.org/10.1016/j.rser.2018.09.024

Bansah KJ, Suglo RS (2016) Sewage treatment by waste stabilization pond systems. J Energy Nat Resource Manag 3(1):8–13. https://doi.org/10.26796/jenrm.v3i1.82

Bateganya N, Nakalanzi D, Babu M, Hein T (2015) Buffering municipal wastewater pollution using urban wetlands in sub-Saharan Africa: a case of Masaka municipality, Uganda. Environ Technol 36(17):2149–2160. https://doi.org/10.1080/09593330.2015.1023363

Bedair H, Alghariani M, Omar E, Anibaba Q, Remon M, Bornman C et al (2023) Global warming status in the African Continent: sources, challenges, policies, and future direction. Int J Environ Res 17:45. https://doi.org/10.1007/s41742-023-00534-w

Bhargava D (2016) Physicochemical wastewater treatment technologies: an overview. Int J Sci Res Educ. https://doi.org/10.18535/ijsre/v4i05.05

Bidira F, Yesuf M, Schaefer N, Friedle M, Alemayehu E (2024) Review on dry toilet and management: brown water (feces) characteristics, composition and management. Environ Health Eng Manage J 11(3):371–384. https://doi.org/10.34172/ehem.2024.36

Blum C, Verdaguer M, Monclús H, Poch M (2020) A new optimization model for wastewater treatment planning with a temporal component. Process Saf Environ Prot 136:157–168. https://doi.org/10.1016/j.psep.2019.12.034

Capodaglio A (2021) Fit-for-purpose urban wastewater reuse: analysis of issues and available technologies for sustainable multiple barrier approaches. Crit Rev Environ Sci Technol 51(15):1619–1666. https://doi.org/10.1080/10643389.2020.1763231

Compaore C, Maiga Y, Nikiema M, Mien O, Nagalo I, Panandtigri H et al (2023) Constructed wetland technology for the treatment and reuse of urban household greywater under conditions of Africa's Sahel region. Water Supply 23(6):2505–2516. https://doi.org/10.2166/ws.2023.121

Crini G, Lichtfouse E (2019) Advantages and disadvantages of techniques used for wastewater treatment. Environ Chem Lett 17(1):145–155. https://doi.org/10.1007/s10311-018-0785-9

Dangui K, Jia S (2022) Water infrastructure performance in Sub-Saharan Africa: an investigation of the drivers and impact on economic growth. Water 14(21):3522. https://doi.org/10.3390/w14213522

Dinka M, Nyika J (2024) SDG 6 progress analyses in sub-Saharan Africa from 2015–2020: the need for urgent action. Discov Water 4:39. https://doi.org/10.1007/s43832-024-00099-5

Drechsel P, Bartram J, Qadir M, Medlicott K (2024) The challenge of supporting and monitoring safe wastewater use in agriculture in LMIC. NPJ Clean Water 6(7):1–3. https://doi.org/10.1038/s41545-024-00364-z

Drechsel P, Danso G, Qadir M (2015) Wastewater use in agriculture: challenges in assessing costs and benefits. In: Wastewater. Springer, pp 139–152. https://doi.org/10.1007/978-94-017-9545-6

Dutta D, Arya S, Kumar S (2021) Industrial wastewater treatment: current trends, bot-tlenecks, and best practices. Chemosphere 285:131245. https://doi.org/10.1016/j.chemosphere.2021.131245

Ebissa G, Fetene A, Desta H (2024) Study on quality of treated wastewater for urban agriculture use in Addis Ababa, Ethiopia. City Environ Interact 24:100157. https://doi.org/10.1016/j.cacint.2024.100157

Edokpayi J, Odiyo J, Popoola O, Msagati T (2021) Evaluation of contaminants removal by waste stabilization ponds: a case study of Siloam WSPs in Vhembe district, South Africa. Heliyon 7(2):e06207. https://doi.org/10.1016/j.heliyon.2021.e06207

Edokpayi J, Odiyo J, Durowoju O (2017) Impact of wastewater on surface water quality in developing countries: a case study of South Africa. Water Qual 401–416

Energizing Development (2024) African biodigester component—assessment of the potential for small-scale and medium-scale biodigesters in Kenya. Deutsche Gesellschaft für Internationale Zusammenarbeit (GIZ) GmbH, Eschborn, Germany

Ezugbe E, Rathilal S (2020) Membrane technologies in wastewater treatment: a review. Membranes 10(5):89. https://doi.org/10.3390/membranes10050089

Figoli A, Nikiema J, Weissenbacher N, Langergraber G, Marrot B, Moulin P (2014) Wastewater treatment practices in Africa—experiences from seven countries in Waterbiotech project. Biotechnology for sustainable water supply in Africa

George-Williams H, Hunt D, Rogers C (2024) Sustainable water infrastructure: visions and options for Sub-Saharan Africa. Sustainability 16(4):1592. https://doi.org/10.3390/su16041592

Gomez-Sanabria A, Zusman E, Höglund-Isaksson L, Klimont Z, Lee S, Akahoshi K et al (2020) Sustainable wastewater management in Indonesia's fish processing industry: bringing governance into scenario analysis. J Environ Manag 275:111241. https://doi.org/10.1016/j.jenvman.2020.111241

Gu Y, Li Y, Li X, Luo P, Wang H, Wang X (2017) Energy self-sufficient wastewater treatment plants: feasibilities and challenges. Energy Procedia 105:3741–3751. https://doi.org/10.1016/j.egypro.2017.03.868

References

Gupta C, Prakash D, Gupta S (2018). Microbes: "A Tribute" to clean environment. In: Paradigms in pollution prevention. SpringerBriefs in Environmental Science. Springer, Cham. https://doi.org/10.1007/978-3-319-58415-7_2

IEA (2015) Sustainable biogas production in municipal wastewater treatment plants. Paris, France. Retrieved from: https://www.ieabioenergy.com/blog/publications/sustainable-biogas-production-inmunicipal-wastewater-treatment-plants/. Accessed on 14 Nov 2024

Ion Exchange (2023) Revolutionizing wastewater treatment in South Africa with membrane bioreactor (MBR) technology. Retrieved from: https://za.ionexchangeglobal.com/revolutionizing-wastewater-treatment-in-south-africa-with-membrane-bioreactor-mbr-technology/. Accessed on 15 Nov 2024

Islam M, Awual M, Angove M (2019) A review on nickel (II) adsorption in single and binary component systems and future path. J Environ Chem Eng 7(5):103305. https://doi.org/10.1016/j.jece.2019.103305

Islam M, Saeed T, Majed N (2022) Role of constructed wetlands in mitigating the challenges of industrial growth and climate change impacts in the context of developing countries. Front Environ Sci 10:1065555. https://doi.org/10.3389/fenvs.2022.1065555

Jones E, Van Vliet M, Qadir M, Bierkens M (2021) Country-level and gridded estimates of wastewater production, collection, treatment and reuse. Earth Syst Sci Data 13(2):237–254. https://doi.org/10.5194/essd-13-237-2021

Kilingo F, Bernard Z, Hong–Bin C (2021) The analysis of wastewater treatment system efficiencies in Kenya: a review paper. Int J Sci Res Publ 11(5):204–215. https://doi.org/10.29322/IJSRP.11.05.2021.p11322

Kim Y, Yoo K, Kim M, Han I, Lee M, Kang B, et al (2019) The capacity of wastewater treatment plants drives bacterial community structure and its assembly. Sci Rep 9(1):1–9. https://doi.org/10.1038/s41598-019-50952

Kjellen M (2018) Wastewater governance and the local, regional and global environments. Water Altern 11(2):219–237

Kubba S (2012) Handbook of green building design and construction. Elsevier INC. https://doi.org/10.1016/C2009-0-64483-4

Lerebours A, Scott R, Sansom K, Kayaga S (2021) Regulating sanitation services in Sub-Saharan Africa: an overview of the regulation of emptying and transport of fecal sludge in 20 cities and its implementation. Utilities Policy 73:101315. https://doi.org/10.1016/j.jup.2021.101315

Maiga Y, Compaore C, Kone M, Sassou S, Some H, Sawadogo M et al (2024) Development of a constructed wetland for greywater treatment for reuse in arid regions: case study in rural Burkina Faso. Water 16(13):1927. https://doi.org/10.3390/w16131927

Makopondo R, Rotich L, Kamau C (2020) Potential use and challenges of constructed wetlands for wastewater treatment and conservation in game lodges and resorts in Kenya. Scientific World J 1–9. https://doi.org/10.1155/2020/9184192

Mara D (2017) The elimination of open defecation and its adverse health effects: a moral imperative for governments and development professionals. J Water Sanit Hyg Dev 7(1):112. https://doi.org/10.2166/washdev.2017.027

Mariwah S, Drangert J, Adams E (2022) The potential of composting toilets in addressing the challenges of fecal sludge management in community-led total sanitation (CLTS). Glob Public Health 17(12):1–14. https://doi.org/10.1080/17441692.2022.2111453

Maryam B, Buyukgungor H (2019) Wastewater reclamation and reuse trends in Turkey: opportunities and challenges. J. Water Process Eng. 30:100501. https://doi.org/10.1016/j.jwpe.2017.10.001

Mehrizi E, Ebrahimi A, Saadati H, Zahedi A, Ghorbanian M, Soltanizadeh Z, Salemi K (2023) Investigating the effectiveness of anaerobic digestion in the treatmen of sugarcane industry wastewater: a systematic review and meta-analysis. Case Stud Chem Environ Eng 8:100414. https://doi.org/10.1016/j.cscee.2023.100414

Mensah J, Silva A, Santos I, Ribeiro N, Gbedjinou M, Nago V et al (2021) Assessment of electricity generation from biogas in Benin from energy and economic viability perspectives. Renew Energy 163:613–624. https://doi.org/10.1016/j.renene.2020.09.014

Mi A, Hi O, Ishaq A, Enokela P, Oluwaseun D, Chikezie A (2022) Performance evaluation of waste stabilization pond for treatment of wastewater from a tertiary institution campus located in Jos North local government area, Plateau state, Nigeria. J Appl Sci Environ Manage 26(9):1523–1528. https://doi.org/10.4314/jasem.v26i9.10

Miarov O, Tal A, Avisar D (2020) A critical evaluation of comparative regulatory strategies for monitoring pharmaceuticals in recycled wastewater. J Environ Manag 254:109794. https://doi.org/10.1016/j.jenvman.2019.109794

Mishra S, Kumar R, Kumar M (2023) Use of treated sewage or wastewater as an irrigation water for agricultural purposes-environmental, health and economic impacts. Total Environ Res Theme 6:100051. https://doi.org/10.1016/j.totert.2023.100051

Muriuki C, Home P, Raude J, Ngumba E, Munala G, Kairigo P et al (2020) Occurrence, distribution and risk assessment of pharmaceuticals in wastewater and open surface drains of peri-urban areas: case study of Juja town, Kenya. Environ Poll 267:115503. https://doi.org/10.1016/j.envpol.2020.115503

Musa M, Idrus S (2021) Physical and biological treatment technologies of slaughter-house wastewater: a review. Sustainability 13(9):4656. https://doi.org/10.3390/su13094656

Nancharaiah Y, Reddy G (2018) Aerobic granular sludge technology: mechanisms of granulation and biotechnological applications. Bioresour Technol 247:1128–1143. https://doi.org/10.1016/j.biortech.2017.09.131

Nansubuga I, Bonadda N, Verstraete W, Rabaey K (2016) A review of sustainable sanitation systems in Africa. Rev Environ Sci Biotechnol 15:465–478. https://doi.org/10.1007/s11157-016-9400-3

Nwokediegwu Z, Dada M, Daraojimba O, Oliha J, Majemite M, Obaigbena A (2024) A review of advanced wastewater treatment technologie s: USA vs Africa. Int J Sci Res Arch 11(01):333–340. https://doi.org/10.30574/ijsra.2024.11.1.0071

Nwosu A, Chukwueloka H (2020) A review of solid waste management strategies in Nigeria. J Environ Earth Sci 10(6):132–143

Obaideen K, Shehata N, Sayed E, Ali M, Mahmoud M, Olabi A (2023) The role of wastewater treatment in achieving sustainable development goals (SDGs) and sustainability guideline. Energy Nexus 7:100112. https://doi.org/10.1016/j.nexus.2022.100112

Oberlin A (2018) Performance evaluation of waste stabilization ponds in Mwanza city, Tanzania. IOSR J Eng 8(4):73–80

Odey E, Abo B, Giwa A, Li Z (2019) Fecal sludge management: insights from selected cities in Sub-Saharan Africa. Arch Environ Prot 45(2):50–57. https://doi.org/10.24425/aep.2019.127984

Okeyo A, Nontongana N, Fadare T, Okoh A (2018) Vibrio species in wastewater final effluents and receiving watershed in South Africa: implications for public health. Int J Environ Res Public Health 15(6):1266. https://doi.org/10.3390/ijerph15061266

Oloruntoba O, Alabi T (2019) Domestic wastewater reclamation and reuse in Nigeria: a case study of some selected treatment plants in Abuja and Lagos. J Future Eng Technol 15(1):1–10. https://doi.org/10.26634/jfet.15.1.14953

Omohwovo E (2024) Wastewater management in Africa: challenges and recommendations. Environ Health Insights 18:1–6. https://doi.org/10.1177/11786302241289681

Onifade A (2024) Wastewater infrastructure in Nigeria and the USA: a tale of two nations and their pathway to sustainable development. Int J Multidispl Res Growth Eval 5(4):179–184. https://doi.org/10.54660/.IJMRGE.2024.5.4.179-184

Onu M, Ayeleru O, Oboirien B, Olubambi P (2023) Challenges of wastewater generation and management in sub-Saharan Africa: a review. Environ Chall 11:100686. https://doi.org/10.1016/j.envc.2023.100686

Othman N, Alias N, Fuzil N, Marpani F, Zaham M, Shahruddin M et al (2021) A review on the use of membrane technology systems in developing countries. Membranes 12(1):30. https://doi.org/10.3390/membranes12010030

References

Ravina M, Galletta S, Dagbetin A, Kamaleldin O, Mng'ombe M, Mnyenyembe L, et al (2021) Urban wastewater treatment in African countries: evidence from the hydroaid initiative. Sustainability 13: 12828. https://doi.org/10.3390/su132212828

Rugaimukamu Q, Wang H, Huang R, Xie L (2022) Wastewater treatment needs more attention. Nature Africa. https://doi.org/10.1038/d44148-022-00071-2

Sackey L, Koomson J, Kumi R, Hayford A, Kayoung P (2023) Assessing the quality of sewage sludge: case study of the Kumasi wastewater treatment plant. Heliyon 9(9):e19550. https://doi.org/10.1016/j.heliyon.2023.e19550

Shyaka E, Niyonzima E (2019) Sustainable design of wastewater treatment plant in Kibiligi Quater. Int J Appl Eng Res 14(22):4105–4111

Silva J (2023) Wastewater treatment and reuse for sustainable water resources management: a systematic literature review. Sustainability 15(14):10940. https://doi.org/10.3390/su151410940

Singh R, Pareek N, Kumar R, Vivekanand V (2022) Anaerobic biodigesters for the treatment of high-strength wastewater. In: Meghvansi M, Goel A (eds) Anaerobic biodigesters for human waste treatment. environmental and microbial biotechnology. Springer, Singapore. https://doi.org/10.1007/978-981-19-4921-0_5

Sinn J, Lackner S (2020) Enhancement of overloaded waste stabilization ponds using different pretreatment technologies: a comparative study from Namibia. J Water Reuse Desalin 10(4):500–512

Sinn J, Agrawal S, Orschler L, Lackner S (2022) Characterization and evaluation of waste stabilization pond systems in Namibia. H2Open J 5(2):365–378. https://doi.org/10.2166/h2oj.2022.004

Steyn M, Walters C, Mathye S, Ndlela L, Thwala M, Banoo I, et al (2021) Atlas of industrial wastewater reuse potential in South Africa. CSIR, South Africa

Tetteh E, Rathilal S, Chetty M, Armah E, Asante-Sackey D (2019) Treatment of water and wastewater for reuse and energy generation-emerging technologies. In: Water and wastewater treatment. IntechOpen, pp 53–80. https://doi.org/10.5772/intechopen.84474

Thebe T, Mangore E (2012) Wastewater production, treatment, and use in Zimbabwe. Retrieved from: Microsoft Word—Format_Country Reports for capacity development project on wastewater Zimbabwe. Accessed on 14 Nov 2024

Tiwari N, Awasthi S (2022) Sewage treatment and reuse by aerobic and anaerobic digestion and physicochemical post-treatment. In: Shah M, Couto S, Shah N, Banerjee R (eds) Development in wastewater treatment research and processes. Elsevier. https://doi.org/10.1016/B978-0-323-85584-6.00023-6

Torrens-Armengol A, de la Varga D, Khafor Ndiaye A, Folch Sánchez M, Coly A (2020) Municipal wastewater treatment and reuse for agriculture in Senegal. Water 12(11):3139. https://doi.org/10.3390/w12113139

UN-Habitat (2023) Global report on sanitation and wastewater management in cities and human settlements. Nairobi, Kenya

UN Habitat and WHO (2021) Progress on wastewater treatment – Global status and acceleration needs for SDG indicator 6.3.1. United Nations Human Settlements Program (UN-Habitat) and World Health Organization (WHO), Geneva

United Nations Environment Program, UNEP (2023) Wastewater—turning problem to solution. A UNEP Rapid Response Assessment. Nairobi, Kenya. https://doi.org/10.59117/20.500.11822/43142

United States Environmental Protection Agency, USEPA (2000) Wastewater technology fact sheet trickling filters. EPA 832-F-00-014, Washington DC, USA

UN-Water (2024) Sub-Saharan Africa. Retrieved from: Region I SDG 6 Data. Accessed on 5 Nov 2024

Waly M, Ahmed T, Abunada Z, Mickovski S, Thomson C (2022) Constructed wetland for sustainable and low-cost wastewater treatment: review article. Land 11: 1388. https://doi.org/10.3390/land11091388

Wang H, Wang T, Zhang B, Li F, Toure B, Omosa I et al (2014) Water and wastewater treatment in Africa—current practices and challenges. Clean: Soil, Air, Water 42(8):1029–1035. https://doi.org/10.1002/clen.201300208

WHO (2024) Improved sanitation facilities and drinking water sources. Retrieved from: https://www.who.int/data/nutrition/nlis/info/improved-sanitation-facilities-and-drinking-water-sources#:~:text=Unimproved%20water%20sources%20include%20unprotected,and%20tanker%20truck%2Dprovided%20water. Accessed on 11 Nov 2024

Wilsenach J, Burke L, Rabede B, Mashego M, Stone W, Mouton M (2013) Denirification in low loaded trickling filters: a case study of the historic trickling filters at the Daspoort wastewater treatment works. Water Research Commission, Pretoria, South Africa

World Bank (2018) 2017 Private Participation in Infrastructure (PPI): Annual Report. The World Bank

World Bank (2019) 2018 Private Participation in Infrastructure (PPI) Annual Re-port. World Bank Retrieved from https://ppi.worldbank.org/content/dam/PPI/documents/PPI_2018_AnnualReport.pdf. Accessed on 11 Nov 2024

WWAP (United Nations World Water Assessment Program) (2017) The United Nations World Water Development Report 2017. Wastewater: The Untapped Resource. Paris, UNESCO

Yadav B, Chavan S, Tyagi R, Drogui P (2022) Occurrence, fate and persistence of per- and polyfluoroalkyl substances (PFASs) during municipal sludge treatment. In: Pilli S, Bhunia P, Tyagi V, Tyagi R, Wong J, Pandey A (eds) Current developments in biotechnology and bioengineering. Elsevier. https://doi.org/10.1016/B978-0-323-99906-9.00005-X

Zhang D, Jinadasa K, Gersberg R, Liu Y, Ng W, Tan S (2014) Application of constructed wetlands for wastewater treatment in developing countries- a review of recent development (2000–2013). J Environ Manage 141:116–131. https://doi.org/10.1016/j.jenvman.2014.03.015

Chapter 4
Challenges Facing Wastewater Management in SSA

Abstract This chapter focused on the challenges that Sub-Saharan Africa (SSA) region faces in its efforts to manage wastewater (WW). Findings showed that WW management in most of the region is flawed due to infrastructural limitations that make effective operations and management of effluents difficult to realize. Treatment plants are not fully functional; they are not able to collect generated effluents effectively; they suffer from power shortages, they have no competent operators and techniques applied cannot remediate effluent pollutants with high removal efficacy. The situation is made worse by the use of conventional treatment technologies, the lack of equipment to assay for effluents prior to and after treatment, corruption during the hiring of plant operators and the non-adherence to predefined standards of WW treatment. Apparent political unwillingness to fund the WW sector and the low social acceptance to the use of treated WW also crippled efforts to manage the resource sustainably. To reverse the current situation, there is a need to revise the existent institutional, legal and governance frameworks for WW management in specific SSA countries in order to infuse more resources, community awareness and desirable attitudes for sustainable management and valorization of effluents.

4.1 Introduction

Wastewater (WW) management is a growing worldwide challenge whereby only 55% of the generated WW is treated while the rest is released and reused in its raw form (UN-Water 2022). The figure is even higher according to Lin et al. (2022) who stated that more than 80% of domestic and industrial WW is discharged without any form of treatment. Of the treated WW, only 22% is reused in all considered countries. Upper-middle income and high-income countries reused 37 and 52%, respectively of the WW share due to the available supporting regulations and the treatment capacities in such countries (Jones et al. 2021). In lower-income countries, WW treatment hardly occurs and such effluents are released to the environment including freshwater bodies without any treatment as a result of limited sanitation and functional sewerage systems (Jones et al. 2021; Lin et al. 2022; Amin et al. 2024).

© The Author(s), under exclusive license to Springer Nature Switzerland AG 2025
J. Nyika and M. O. Dinka, *The Silent Wastewater Problem in Sub-Saharan Africa*,
SpringerBriefs in Water Science and Technology,
https://doi.org/10.1007/978-3-031-90143-0_4

Reusing untreated WW poses a risk of infectious disease transmission to humans in addition to adversely affecting ecosystems and natural resources (Qadir et al. 2020; Amin et al. 2024).

Even with the adverse effects associated with the reuse of untreated WW, the trend is growing especially downstream areas of urban areas where informal agricultural activities occur (Drechsel et al. 2024). The Food and Agriculture Organization (FAO) reported that irrigation using partially treated and untreated WW contributed to 10% of land under irrigation, which is approximately 20 million hectares in more than 50 countries (Winpenny et al. 2013). Thebo et al. (2017) reported that more than 39 million hectares of land in downstream areas of urban centers across the world are irrigated with untreated or partially treated WW. The increased use of the resource is driven by factors such as the limited availability of freshwater bodies, the apparent water scarcity and limited regulatory provisions on WW reuse.

Most of the unplanned WW reuse for irrigation occurs following the dilution of the resource once it is mixed with fresh surface water such as in lakes, streams and rivers (Drechsel et al. 2024). However, the dilution effect is not adequate to reduce the risk of pathogenic transmissions and as such, reuse of WW in such cases is mostly unsafe. In Sub-Saharan Africa (SSA) region whereby urban and peri-urban agriculture is growing due to the expansion of the informal sector, most vegetables are grown via WW irrigation (Thebo et al. 2017; Nyika 2022). The magnitude of the resultant effects of untreated and partially treated WW reuse in the region could worsen considering that water quality is hardly monitored and assessment and monitoring of its extent and associated risk is hardly done (Drechsel et al. 2024). To tackle this challenge, it is imperative to understand the challenges that hinder scientific WW management in the region to enable better planning and affirmative action in the future. Drechsel et al. (2022) noted that the shift from unplanned to planned, informal to formal and unsafe to safe WW reuse in many countries particularly those of the global south such as SSA require better understanding of the existent WW system and its inherent challenges in order to formulate viable and sustainable solutions moving forward. Therefore, the aim of this chapter is to explore the challenges that SSA encounters in its attempts to manage WW using named case examples in the region.

4.1.1 Challenges in Wastewater Management in SSA

The lack of sanitation in SSA region has resulted to the lack of safe water supply for more than 300 million people in the region and the wide spread of water-borne diseases (Yang et al. 2020). The diseases are a result of the discharge of raw WW into the environment. The WW management problem is further exacerbated by the need for effective and acceptable WW treatment (Dos Santos et al. 2017). Although the region is making efforts to continuously provide quality water for its population, the measures are not commensurate with expanding urbanization, industrialization, population rise and the increase WW production tendencies (Nyika and Dinka 2023). Consequently, water resources in the region are under pressure and the quality is on a declining trend. The challenges associated with poor management of WW in the region are discussed in the following subtopics.

4.1.2 Insufficient WW Infrastructure

Residents in SSA nations do not have adequate WW services and as such collection, treatment and reuse of the resource hardly occurs. Despite the increase in WW production levels in the region, collection, treatment and reuse rates in most countries with exception of Southern Africa and Nigeria was less than 10% (UNU-INWEH 2024). In most recent statistics, only 20% of the total generated WW in the region was safely treated while the rest was discharged to the environment in its raw form (UN-Water 2024).

In areas where infrastructure exists, it is either inadequate or obsolete. In Ethiopia's capital, Addis Ababa, the Kaliti WW treatment plant designed in 1982 to serve 50,000 people was found to serve more than 200,000 people 30 years later despite zero modifications in its design (Ravina et al. 2021). The plant is currently overloaded despite minimal rehabilitation. Most of the residents in Addis Ababa have no connection to sewer lines in their households and as such only less than 3% of the generated WW ends up in the treatment facilities as Abiye et al. (2009) noted. This observation is also common in many SSA countries (Omohwovo 2024).

In another case in Kisumu, Kenya, the only three WW pumping stations available in the city were reported to be broken down due to sewage overflow in the available manholes. As a result, most of the generated WW was discharged directly to Lake Victoria, which is a freshwater resource (Parkman et al. 2008). In Kigali, Rwanda, there is no centralized WW treatment plant and there is no central sewer network and as such, majority of the population use pit latrines and soak-away pits to manage generated sewage (Nzitonda 2023). The situation is similar in Mali's capital, Bamako where more than 80% of the population relies on on-site WW treatment due to lack of treatment facilities (World Bank Group 2022).

In Ghana's urban areas, 4–5% of the population is underserved by WW services since the sewerage networks and treatment plants are infrequently dysfunctional. The tendency results to most of the effluent ending up in streams, storm water gutters and freshwater bodies where it is contaminated before use (Gyampo and Nutsukpo 2012). Owing to these infrastructural limitations, only 10% of domestic WW is collected to the sewer networks while 20 and 70% of rural and urban population has access to toilets (Gyampo and Nutsukpo 2012). Most industries located in Lagos and Abuja, urban Nigeria have no localized effluent treatment plants and are not connected to sewer systems and as such, they discharge raw WW to freshwater bodies (Kayode et al. 2018). In Zambia's Lusaka, most of the residents have no access to sanitation and as such, sewage networks are limited and fecal sludge is poorly management (Ghesquiere et al. 2024). Consequently, environmental degradation and the prevalence to waterborne diseases has become common and significant economic losses have been incurred at national level. In Zimbabwe, high prevalence of cholera in 2008 was associated with sanitation and infrastructural limitations since WW treatment plants could not adequately handle the effluents (Amarachi et al. 2023).

WW is of diverse nature in that it comprises of chemical (nutrients, pharmaceuticals and personal care products, poly- and perfluoroalkyl substances, biocides, heavy metals, dyes, radionuclides and transformation products) and non-chemical (microplastics, pathogens and nanoparticles) constituents (Villarin and Merel 2020). The mixture of the constituents makes WW toxic and assessment of the acute and chronic toxicity associated with them is challenging especially if frequent monitoring is not done as is the case in SSA region (Villarin and Merel 2020).

In SSA region, water quality monitoring is poorly done and most laboratories are not equipped to assay and quantify the constituents of WW so that appropriate remediation measures can be taken (Re et al. 2011). Existent laboratories especially those located in WW treatment plants have no monitoring tools in addition to lacking qualified personnel (Omohwovo 2024). In SSA region, monitoring of WW and its constituents is hampered by lack of capacity and poor equipping of facilities that are charged with such assaying (AfDB, UNEP and GRID-Arendal 2020; UN-Habitat 2023). As such, the inadequacies in monitoring WW makes controlled discharge difficult to realize. In Nairobi, Kenya, only a few parameters such as alkalinity, pH and turbidity can be monitored in river water, which is suspected to be contaminated with WW due to lack of instrumentation (Wang et al. 2014). In Uganda where illegal industrial effluent discharge is known to the environmental management regulatory agencies, follow up monitoring and assessment as well as enforcement of existent laws is difficult because an effective system to do so does not exist (Wang et al. 2014).

4.1.3 Financial Unsustainability

The already financially challenged SSA countries experience difficulties prioritizing for WW management and allocating resources for it. Without adequate finances, procuring equipment for WW analyses, refurbishing and expanding the sewer networks and treatment technologies and capacity is difficult (Wang et al. 2014). Additionally, setting up sanitation facilities to manage WW is nearly impossible and for this reason, the handling of effluent in the region uses conventional and on-site treatment systems, which in most cases are ineffective (Drechsel et al. 2024). Capital expenditure to build WW infrastructure, operate and maintain it is also limited in SSA region hence planning, budgeting and financial reporting on WW investments is hardly done (AfDB, UNEP and GRID-Arendal 2020). Lack of optimization and modernization of WW treatment processes hinders efficiency and effectiveness in the management processes and in taking affirmative action for the already strained WW sector (Omohwovo 2024).

In Kenya's capital, Nairobi, the jar tester dating method to assay treated WW is still being used since the early 1900s due to limited financial capacity to use tech-savvy techniques of assaying for pollutants in effluents (Wang et al. 2014). Cameroon's

4.1 Introduction

ability to handle WW in rural and urban areas has been hindered by limited financial capacity (Omohwovo 2024). In South Africa, the sources of funding to enhance sanitation infrastructure for struggling municipalities to manage WW are limited and it is difficult to treat the resource prior to its discharge and subsequent reuse safely (Khuzwayo and Chirwa 2020). The costs associated with management of WW in the country are also difficult to recover with financial constraints and as such, effluent recycling and enhancement of sewer networks to improve coverage of collection services are difficult to meet (Khuzwayo and Chirwa 2020). In urban areas of Zimbabwe (Graham Sustainability Institute University of Michigan 2024), Botswana (Mmereki et al. 2014), and Gabon (Onu et al. 2023) like most SSA nations, the lack of local and international financial resources to overhaul and upgrade waste collection and treatment facilities and technologies makes management and recovery of the resource problematic. A similar pattern was established in Benin, Malawi and Ethiopia and attributed to the financial limitations of the countries such that funding to improve the WW sector is not a priority (Ravina et al. 2021).

4.1.4 Limited Technical and Human Capacity

The lack of adequate training and skills on WW management in SSA region is a drawback to the resource's efficient handling and treatment. According to UNESCO-UNEVOC (2012), lower income countries of SSA do not have enough professionals and technicians to plan, design, operate and maintain WW collection and treatment systems. Similarly, Wang et al. (2014), pointed out that the lack of expertise to operate and maintain WW systems in SSA is a hindrance to sustainable operations in the sector. The World Water Assessment Program, WWAP (2017) also reported that the lack of technical and human expertise was a limitation to managing WW efficiently in Tanzania among other SSA nations. In South Africa, WW management was crippled by the lack of proper education and training for treatment process controllers and plant operators, poor management and administrative skills and inadequate research on best practices in WW management (Khuzwayo and Chirwa 2020). In Kenyan resorts and game lodges, the lack of expertise to operate and maintain constructed wetlands was associated with inadequate pollutant remediation from WW (Makopondo et al. 2020). The lack of skilled human capacity to manage sewage and effluents in SSA countries of Cameroon, Rwanda and Zambia was attributable to inadequate treatment and extensive pollution of freshwater system (AfDB, UNEP and GRID-Arendal 2020).

4.1.5 Regulatory and Governances Limitations

In SSA region, WW management and sanitation challenges are many. They include low priority accorded to effluent management and improvement to sanitation facilities, limited data on collection, treatment and reuse of WW, poor communication and coordination among stakeholders and poor integration of sanitation, hygiene, water and WW issues (Ekane et al. 2014). Additional governance issues include gaps in policy coordination and research on WW, sociopolitical unwillingness to spend in the sector, poor enforcement of decentralized solutions in the sector, insufficient human resources capacity, non-commitment to implement tech-savvy technology to manage WW and inappropriate legal and institutional frameworks (Szanto et al. 2012; UN-Water 2012; Ekane et al. 2012; Ekane and Gill 2013). A disconnect between current practice and sanitation and hygiene regulations hence insufficient focus on WW management in general is also apparent (Szanto et al. 2012; UN-Water 2012; Ekane et al. 2012; Ekane and Gill 2013).

The governance and regulation challenges in SSA are encountered at household, local and regional levels of governance. In the region, governments and politicians do not prioritize WW and water management because it does not help them get votes (Wang et al. 2014). The apparent low priority in the sector has led to the neglect in building WW infrastructure and setting up policies to regulate its safe discharge to the environment. The lack of public participation of stakeholders in the WW sector is also ailing the management of the resource. In SSA region, WW management decisions are made without the involvement of the public who in most cases do not have adequate awareness of the need to protect freshwater resources, the environment and human health by enhancing safe effluent discharge (Wang et al. 2014). Most SSA countries adopt guidelines by WHO on WW treatment and quality standards prior to discharge because they do not have their own regulatory framework. However, even the WHO guidelines are not adequate to improve the sector due to poor monitoring tools, lack of equipped laboratories, limited water and WW quality control and lack of a strong institutional framework to manage sewage (Wang et al. 2014).

In Uganda, for instance, politicians have no commitments in improving the WW sector and do not fulfill the promises they make to improve the sector through strong policy enactment and enforcement in collaboration with local leaders and citizens once they are elected (Quin et al. 2011). The lack of stringent regulation and rules to control raw WW discharge and a poorly coordinated plan to manage industrial effluent and sewage in Kenya, Uganda and Tanzania of East Africa is attributed to the extensive pollution of Lake Victoria (Wang et al. 2012a). In informal settlements of Nairobi, Kenya, WW management was reported to be neglected and information and data on generation rates from all sources were limited, and residents had no access to collection, treatment and disposal services (Kilingo et al. 2022). In Abuja, Nigeria, city regulators neglected WW management and for this reason, 55% of residents used the informal sector to supplement government services of collection and treatment while 68% invested personal resources to improve individual WW management approaches on-site (Abubakar 2017).

At city and national level in Kenya and Malawi, on-site sanitation and use of pit latrines is prevalent and has resulted to fecal sludge management (Yesaya and Tilley 2021). The authors attributed the problem to low precedence to effective sanitation practices, the lack of strong policies to enforce WW management and the non-commitment by the government to adopt advanced technologies to manage the sludge. Municipalities in different provinces of South Africa (Gauteng, Western

4.1 Introduction 73

Cape, Mpumalanga, Northern Cape, Limpopo, Free State and KwaZulu-Natal) had neglected their WW management responsibilities (Herbig 2019). The municipalities were non-compliant to WW regulations in their jurisdictions, treated water below the predefined quality standards by South African laws, did not conduct laboratory analysis after WW treatment scientifically and employed treatment plant operators on the basis of corruption (Herbig 2019). Ultimately, these regulatory and governance misdoings resulted to deteriorating WW management and a greater etiology to human health and the environment at local, regional and national levels in all SSA countries.

4.1.6 Public Awareness and Education

Education and community awareness on the need to manage WW is imperative in safeguarding the health of living things including humans and environmental health (Amarachi et al. 2023). In developing countries particularly those of SSA region, the lack of education on the need to protect freshwater systems and the environment at large is attributable to poor uptake of sustainable WW management approaches (Massoud et al. 2009). Due to the lack of education and community awareness, many SSA residents consider payment for sanitation and effluent treatment high in addition to undermining the use of better hygiene measures such as the use of pit latrines (Dittmer 2009; Massoud et al. 2009). Limited education on hygiene and sanitation and its effects on health in West African countries of Nigeria, Mali, Ghana (Dittmer 2009) and Cameroon (Yongsi 2009) was attributable to the extensive use of on-site WW treatment systems and the unsafe discharge of raw effluents. The illiteracy tendencies results to negative attitudes towards the management of WW and its associated valorization and reuse after treatment (Ali and Gujiba 2024).

With low community awareness and education, WW reuse after treatment is unlikely to be accepted socially (Villarin and Merel 2020). In comparison to non-potable uses of WW such as irrigation, the use of treated effluents for drinking is less likely to be accepted without adequate education and due to the "yuck factor" associated with such recycled water (Smith et al. 2018). In Burkina Faso (Campaore et al. 2024) and Nigeria (Akpan et al. 2020), illiteracy was associated with mismanagement of WW including its discharge in its raw form and also discomfort in using recycled WW especially for potable purposes. Despite the necessity to reuse WW in water-scarce SSA countries, it was reported that social acceptance of such advances was low in South Africa due to limited knowledge and awareness among sewage generators of such uptakes and there importance in alleviating water shortage (Bakare et al. 2016).

4.1.7 Poor Operations and Maintenance

The operation and maintenance of many WW treatment plants in SSA is challenged and in most cases not comparable to the effluents they are receiving. According to Ali and Gujiba (2024), most treatment plants of SSA located in urban areas do not adequately collect, store, transport and treat WW because they are either not working or originally, they were irreversibly polluted and treatment technologies applied are not able to remediate contaminants. The operations of various SSA industries do not accommodate effluent treatment due to cost constraints and unavailability of appropriate technologies (Wang et al. 2014). In events where turbidity of WW is very high or low, waterworks also fail to work especially when the dosages of coagulants are not correct. Such an observation was made in Nairobi, Kenya whereby, the turbidity of water was more than 5000 nephelometric turbidity units (NTU) in rainy seasons due to the mixing of WW with sediments and could reduce to <10 NTU during the dry seasons (Wang et al. 2012b). With such fluctuations, variations in the coagulant dosage are required and as such, WW networks can become blocked due to clogging. The use of algae in biological WW treatment can also result to consumption of coagulants among other chemicals used in effluent management resulting to complications such as fouling, growth of toxic microcystin and the clogging of filters in the sewer networks (Okello et al. 2010; Wang et al. 2014).

Important steps are never taken in SSA countries to treat WW contaminated water before it is used for drinking. Such steps include disinfection, which in most cases is not done before release of sewage to freshwater systems that livestock use for drinking in South Africa (Du Preez et al. 2010) and in Kenya (Du Preez et al. 2011). Effluent from stabilization ponds at the Dandora WW treatment plant was discharged to the Nairobi River (Kenya) where livestock used the water for drinking without any disinfection (Du Preez et al. 2011). Treated WW in Bugolobi WW treatment plant of Uganda was found to have elevated levels of *Escherichia coli* compared to the predefined standards of discharging treated effluents to the environment (Wang et al. 2014). Such sub-standard operations are likely to induce harm to public health, livestock, human and the environment. The lack of pretreatment of WW before releasing them in lagoons and stabilization ponds results to inefficient contaminant remediation due to heavy pollutant concentrations of the influent. Such a tendency was attributable to the low removal efficacy of chemical and biological oxygen demand among other contaminants in treated WW effluents from Dandora treatment plant in Kenya (Li et al. 2011).

Efforts to treat WW in many SSA countries have been hampered by the apparent energy insecurities in the region. According to Wang et al. (2012b, 2014), unreliable power supplies characterized by frequent blackouts prevent effective WW treatment in many SSA countries. In Nigeria, WW treatment is hampered by the lack of consistent power supply (Omohwovo 2024). The functionality of many WW treatment plants in SSA is interrupted by high power costs, power cuts, overloading and compliance failures as reported in South Africa, Burkina Faso, Ghana, and Senegal (Water Aid 2019). In the countries, fecal sludge was therefore not well treated due to the power challenges. Additionally, the operations and maintenance challenges in most WW treatment plants in urban Tanzania and Senegal like most SSA countries were partially or not functional at all (UN-Habitat 2023). Equally, the on-site treatment technologies employed to manage sewage were ineffective to decontaminate it before environmental release (UN-Habitat 2023).

4.1.8 Emerging Contaminants

WW comprises of constituents such as microplastics, radionuclides, mycotoxins, pesticides, volatile organic compounds, pharmaceuticals, polycyclic aromatic hydrocarbons (PAHs) and personal care products all of which are emerging contaminants (ECs) (Amarachi et al. 2023). The ECs have negative effects on freshwater resources, cause biodiversity loss, negative effects on ecosystems and wildlife and result to endocrine disruption, neurotoxicity, carcinogenicity, immunotoxicity, respiratory dysfunction, cardiovascular and metabolic complications (Li et al. 2024). To detect and identify the ECs requires advanced techniques and as such data on their levels in WW are rarely published or its establishment is at the nascent stages.

In SSA region, existent WW treatment systems are not designed to quantify and remediate ECs (Ripanda et al. 2022). Additionally, treatment techniques applied are ineffective and the existent regulatory framework does not include the analysis of the contaminants in routine analysis of WW prior and after treatment. ECs including anti-malarial drugs, steroid hormones, anti-epileptics, antiretrovirals, anitbiotics and non-steroid anti-inflammatories have been found in SSA effluents (Lapworth et al. 2017; Madikizela et al. 2017, 2020; Ngubane et al. 2019). Such ECs identified in effluents and freshwater systems contaminated by WW in various SSA countries are as shown in Table 4.1.

Owing to the infrastructural and technological challenges in the WW sector in SSA, reports on the prevalence of ECs are rare and those that are available are not deemed as adequate in advising regulatory agencies on policy revisions (Ripanda et al. 2022). The results identified in existent studies would need to be validated using advanced laboratory techniques and equipment such as mass spectrometry and high-pressure liquid chromatography. However, the technologies are not available in SSA and as such, ECs are likely to accumulate in the environment and their effects will become grave in the near future (Khalid et al. 2018; Teta et al. 2018; Tapela and Rahube 2019). Immigrants from 19 central and western African countries who travelled to Spain and were previously exposed to ECs had polychlorinated biphenyls (PCBs) in their blood and organochloride pesticides in their serum due to consumption of polluted WW. Therefore, without using advanced approaches to assay and manage WW, SSA stands to bear the adverse effects associated with ECs among other pollutants found in its effluents.

Table 4.1 Emerging contaminants found in effluents and freshwater systems contaminated by WW in SSA (Ripanda et al. 2022)

Country	Identified ECs
Botswana	Antibiotic resistant determinants: penicillin, ampicillin, erythromycin, sulfamethoxazle
Lesotho	PAHs, fly ash, mercury
Namibia	Testosterone, estradiol, PCBs, organic pesticides
Democratic Republic of Congo	Heavy metals and organic pesticides
Congo Brazzaville	Organic pesticides, antibiotics
Gabon	Antibacterial resistance genes
Cameroon	Organic pesticides such as diuron, linuron, chlortoluron and atrazine Pharmaceutical compound including antibiotics
Guinea Bissau	Heavy metals including methyl mercury, lead and arsenic
Cape Verde and Mali	Persistent organic contaminants POCs including endosulfan, diazinon, dacthal, trans-chlordane, PCBs
Burkina Faso and Togo	Organochlorine pesticides including aldrin and endosulfan
Ghana	Antibiotics, pharmaceutical residues
Kenya	Analgesics, antibiotics, antivirals, psychiatric drugs
Uganda and Tanzania	Antibiotics and analgesics
Zambia	Antibiotics and antivirals
Zimbabwe	Androgens, dihydrotestosterone, estradiol
Mozambique and Ethiopia	Antibiotics and analgesics
Malawi	Organic pesticides
South Sudan	Lead
Seychelles	PFASs, PCBs, organochlorine pesticides

4.2 Conclusion

The findings of this chapter show that a number of challenges impede effective WW management in SSA. In most nations of the region, WW infrastructure is dysfunctional characterized by the lack of connection to sewer lines for residents, sewage overflow and discharge to the environment in its raw form, limited sanitation facilities to promote WW management and non-operational or partially operational treatment plants. The situation is further exacerbated by poor financial support to improve the existent infrastructure that is overloaded and/or obsolete. The region has limited experts, laboratories, equipment and tools to assay for complex WW pollutants such as ECs before and after treatment of effluents. Consequently, data collection on generation, collection, reuse and treatment efficacy patterns of WW is hardly collected and affirmative management measures are not implemented. The non-prioritizing of WW management in prevailing politics of the region has barred regulatory, policy and institutional improvements and action plans to manage the resource and enhance community awareness on the need for sustainable resource valorization. Ultimately, the functionality of the WW sector is flawed and demotes energy security, human and environmental health unless effective measures, technological, infrastructural and financial inputs are provided to improve the current state in the region.

References

Abiye T, Sulieman H, Ayalew M (2009) Use of treated wastewater for managed aquifer recharge in highly populated urban centers: a case study in Addis Ababa, Ethiopia. Environ Geol 58(1):55–59. https://doi.org/10.1007/s00254-008-1490-y

Abubakar I (2017) Household response to inadequate sewerage and garbage collection services in Abuja, Nigeria. J Environ Public Health 5314840. https://doi.org/10.1155/2017/5314840

African Development Bank, United Nations Environment Program and GRID-Arendal (2020) Sanitation and wastewater atlas of Africa. AfDB, UNEP and GRID-Arendal. Abidjan, Nairobi and Arendal

Akpan V, Omole D, Bassey D (2020) Assessing the public perceptions of treated wastewater reuse: opportunities and implications for urban communities in developing countries. Heliyon 6(10):e05246. https://doi.org/10.1016/j.heliyon.2020.e05246

Ali A, Gujiba U (2024) Household wastewater management in sub-Saharan Africa: a review. Discov Water 4:6. https://doi.org/10.1007/s43832-024-00060-6

Amarachi A, Amaefule E, Agbai S, Abuka C, Nwaogazie F, Ajamu P et al (2023) Sustainable water and wastewater management: challenges, innovations and lessons from global case studies. Path Sci 9(9):5014–5026. https://doi.org/10.22178/pos.96-11

Amin N, Foster T, Shimki N, Willetts J (2024) Hospital wastewater (HWW) treatment in low- and middle-income countries: a systematic review of microbial treatment efficacy. Sci Total Environ 921:170994. https://doi.org/10.1016/j.scitotenv.2024.170994

Bakare B, Mtsweni S, Rathilal S (2016) A pilot study into public attitudes and perceptions towards greywater reuse in a low cost housing development in Durban, South Africa. J Water Reuse Desal 6(2):345–354. https://doi.org/10.2166/wrd.2015.076

Campaore C, Ouili A, Zongo S, Dabre D, Maiga Y, Mogmenga I, Pale D et al (2024) Assessing greywater characteristics in the Sahel region and perception of the local population on its reuse in agriculture. Heliyon 10(14):e33473. https://doi.org/10.1016/j.heliyon.2024.e33473

Dittmer A (2009) Towards total sanitation socio-cultural barriers and triggers to total sanitation in West Africa. Retrieved from: http://www.wsscc.org. Accessed 20 Nov 2024

Dos Santos S, Adams E, Neville G, Wada Y, De Sherbinin A, Bernhardt E, Adamo S (2017) Urban growth and water access in sub-Saharan Africa: progress, challenges, and emerging research directions. Sci Total Environ 607:497–508. https://doi.org/10.1016/j.scitotenv.2017.06.157

Drechsel P, Qadir M, Galibourg D (2022) The WHO guidelines for safe wastewater use in agriculture: a review of implementation challenges and possible solutions in the global south. Water 14(6):864. https://doi.org/10.3390/w14060864

Drechsel P, Bartram J, Qadir M, Medlicott K (2024) The challenge of supporting and monitoring safe wastewater use in agriculture in LMIC. NPJ Clean Water 6(7):1–3. https://doi.org/10.1038/s41545-024-00364-z

Du Preez M, McGuigan K, Conroy R (2010) Solar disinfection of drinking water in the prevention of dysentery in south African children aged under 5 years: the role of participant motivation. Environ Sci Technol 44(22):8744–8749. https://doi.org/10.1021/es103328j

Du Preez M, Conroy R, Ligondo S, Hennessy J, Elmore-Meegan M, Soita A et al (2011) Randomized intervention study of solar disinfection of drinking water in the prevention of dysentery in Kenyan children aged under 5 years. Environ Sci Technol 45(21):9315–9323. https://doi.org/10.1021/es2018835

Ekane N, Gill T (2013) Sanitation policy and practice in Rwanda: tackling the disconnect. SEI Policy Brief, Stockholm Environment Institute, Stockholm

Ekane N, Noel S, Kjellen M, Fodge M (2012) Sanitation and hygiene: policy, stated briefs and actual practice: a case study in the Burera district in Rwanda. Retrieved from: https://www.sei.org/publications/sanitation-and-hygiene-policy-stated-beliefs-and-actual-practice-a-case-study-in-the-burera-district-rwanda/. Accessed on 20 Nov 2024

Ekane N, Nykvist B, Kjellen M, Noel S, Weitz N (2014) Multi-level sanitation governance: understanding and overcoming challenges in the sanitation sector in Sub-Saharan Africa. Waterlines 33(3):242–256. https://doi.org/10.3362/1756-3488.2014.024

Ghesquiere F, Huang A, Mayumbelo K, Malate K, Wakunuma M (2024) Lessons from Zambia: improving sanitation through planning, awareness-building and collaboration. World Bank Group, Washington DC, USA

Graham Sustainability Institute University of Michigan (2024) Linking research and management for safe and sustainable water supply by drinking water utilities. Retrieved from: https://graham.umich.edu/activity/13209. Accessed on 19 Nov 2024

Gyampo M, Nutsukpo D (2012) Wastewater production, treatment and use in Ghana. 3rd regional workshop "Safe use of wastewater in agriculture", 26–28 September, 2012, Johannesburg, South Africa

Herbig F (2019) Law, criminology & criminal justice | review article talking dirty—effluent and sewage irreverence in South Africa: a conservation crime perspective. Cogent Soc Sci 5:1701359. https://doi.org/10.1080/23311886.2019.1701359

Jones E, Van Vliet M, Qadir M, Bierkens M (2021) Country-level and gridded estimates of wastewater production, collection, treatment and reuse. Earth Syst Sci Data 13:237–254. https://doi.org/10.5194/essd-13-237-2021

Kayode O, Luethi C, Rene E (2018) Management recommendations for improving decentralized wastewater treatment by the food and beverage industries in Nigeria. Environments 5(3):41. https://doi.org/10.3390/environments5030041

Khalid S, Shahid M, Bibi I, Sarwar T, Shah A, Niazi N (2018) A review of environmental contamination and health risk assessment of wastewater use for crop irrigation with a focus on low and high-income countries. Int J Environ Res Pub Health 15:895. https://doi.org/10.3390/ijerph15050895

Khuzwayo Z, Chirwa E (2020) The intricate challenges of delocalised wastewater treatment facilities with regards to water resource management capacity framework in South Africa. Sustain Water Resour Manag 6(6). https://doi.org/10.1007/s40899-020-00367-x

Kilingo F, Bernard Z, Hongbin C (2022) Study of domestic wastewater treatment using *Moringa oleifera* coagulant coupled with vertical flow constructed wetland in Kibera Slum, Kenya. Environ Sci Pollut Res 29:36589–36607. https://doi.org/10.1007/s11356-022-18692-3

Lapworth D, Nkhuwa D, Okotto-Okotto J, Pedley S, Stuart M, Tijani M, Wright J (2017) Urban groundwater quality in sub-Saharan Africa: current status and implications for water security and public health. Hydrogeol J 25(4):1093–1116. https://doi.org/10.1007/s10040-016-1516-6

Li F, Wang H, Mafuta C (2011) Current status and technology demands for water resources and water environment in Africa, in research on water resources of African typical areas. Science Press, Beijing, China

Li X, Shen X, Jiang W, Xi Y, Li S (2024) Comprehensive review of emerging contaminants: detection technologies, environmental impact and management strategies. Ecotoxicol Enxiron Saf 278:116420. https://doi.org/10.1016/j.ecoenv.2024.116420

Lin L, Yang H, Xu X (2022) Effects of water pollution on human health and disease heterogeneity: a review. Front Environ Sci 10:880246. https://doi.org/10.3389/fenvs.2022.880246

Madikizela L, Tavengwa N, Chimuka L (2017) Status of pharmaceuticals in African water bodies: Occurrence, removal and analytical methods. J Environ Manag 193:211–220. https://doi.org/10.1016/j.jenvman.2017.02.022

References

Madikizela L, Ncube S, Chimuka L (2020) Analysis, occurrence and removal of pharmaceuticals in African water resources: A current status. J Environ Manag 253:09741. https://doi.org/10.1016/j.jenvman.2019.109741

Makopondo R, Rotich L, Kamau C (2020) Potential use and challenges of constructed wetlands for wastewater treatment and conservation in game lodges and resorts in Kenya. Scientific World J 1–9. https://doi.org/10.1155/2020/9184192

Massoud M, Tarhini A, Nasr J (2009) Decentralized approaches to wastewater treatment and management: applicability in developing countries. J Environ Manage 90(1):652–659. https://doi.org/10.1016/j.jenvman.2008.07.001

Mmereki D, Li B, Meng L (2014) Hazardous and toxic waste management in Botswana: practices and challenges. Waste Manag Res 32:1158–1168. https://doi.org/10.1177/0734242x14556527

Ngubane Z, Dzwairo B, Moodley B, Stenstrom T, Sokolova E (2019) Quantitative assessment of human health risks from chemical pollution in the uMsunduzi River, South Africa. Environ Sci Pollut Res 30:118013–118024. https://doi.org/10.1007/s11356-023-30534-4

Nyika J, Dinka M (2023) Water challenges in rural and urban Sub-Saharan Africa and their management. Springer Nature, Cham, Switzerland

Nyika J (2022) Wastewater for agricultural production, benefits, risks, and limitations. In: Chatoui H, Merzouki M, Moummou H, Tilaoui M, Saadaoui N, Brhich A (eds) Nutrition and human health. Springer, Cham. https://doi.org/10.1007/978-3-030-93971-7_6

Nzitonda J (2023) Inclusive urban sanitation stories. Towards improving fecal sludge management in Kigali, Rwanda. IWA, London, United Kingdom

Okello W, Ostermaier V, Portmann C, Gademann K, Kurmayer R (2010) Spatial Isolation favors the divergence in microcystin net production by microcystis in Ugandan freshwater lakes. Water Res 44(9):2803–2814. https://doi.org/10.1016/j.watres.2010.02.018omoh

Omohwovo E (2024) Wastewater management in Africa: challenges and recommendations. Environ Health Insights 18:1–6. https://doi.org/10.1177/11786302241289681

Onu M, Ayeleru O, Oboirien B, Olubambi P (2023) Challenges of wastewater generation and management in sub-Saharan Africa: a review. Environ Chall 11:100686. https://doi.org/10.1016/j.envc.2023.100686

Parkman M, Patchett J, Odongo O (2008) Kisumu water supply and sanitation project, long term action plan: water design report, LakeVictoria South Water Services Board, Kisumu, Kenya

Qadir M, Drechsel P, Jiménez Cisneros B, Kim Y, Pramanik A, Mehta P et al (2020) Global and regional potential of wastewater as a water, nutrient and energy source. Nat Resour Forum 44:40–51. https://doi.org/10.1111/1477-8947.12187

Quin A, Balfors B, Kjellen M (2011) How to "walk the talk": the perspectives of sector staff on implementation of the rural water supply program in Uganda. Nat Resour Forum 35(4):269–282. https://doi.org/10.1111/j.1477-8947.2011.01401.x

Ravina M, Galletta S, Dagbetin A, Kamaleldin O, Mng'ombe M, Mnyenyembe L, et al (2021) Urban wastewater treatment in African countries: evidence from the hydroaid initiative. Sustainability 13: 12828. https://doi.org/10.3390/su132212828

Re V, Faye S, Faye A, Faye S, Gaye C, Sacchi E, Zuppi G (2011) Water quality decline in coastal aquifers under anthropic pressure: the case of a suburban area of Dakar (Senegal). Environ Monit Assess 72(1–4):605–622. https://doi.org/10.1007/s10661-010-1359-x

Ripanda A, Rwiza M, Nyanza E, Njau K, Vuai S, Machunda R (2022) A review on contaminants of emerging concern in the environment: a focus on active chemicals in Sub-Saharan Africa. Appl Sci 12(1):56. https://doi.org/10.3390/app12010056

Smith H, Brouwer S, Jeffrey P, Frijns J (2018) Public responses to water reuse- understanding the evidence. J Environ Manag 207:43–50. https://doi.org/10.1016/j.jenvman.2017.11.021

Szanto G, Letema S, Tukahirwa J, Mgana S, Oosterveer P, van Buuren J (2012) Analyzing sanitation characteristics in the urban slums of East Africa. Water Policy 14:613–624. https://doi.org/10.2166/wp.2012.093

Tapela K, Rahube T (2019) Isolation and antibiotic resistance profiles of bacteria from influent, effluent and downstream: a study in Botswana. African J Microbiol Res 13(15):279–289. https://doi.org/10.5897/AJMR2019.9065

Teta C, Holbech BF, Norrgren L, Naik YS (2018) Occurrence of oestrogenic pollutants and widespread feminisation of male tilapia in peri-urban dams in Bulawayo, Zimbabwe. African J Aquati Sci 43(1):1–10. https://doi.org/10.2989/16085914.2017.1423269

Thebo A, Drechsel P, Lambin E, Nelson K (2017) A global, spatially-explicit assessment of irrigated croplands influenced by urban wastewater flows. Environ Res Lett 12:074008. https://doi.org/10.1088/1748-9326/aa75d1

UN-Habitat (2023) Global report on sanitation and wastewater management in cities and human settlements. Nairobi, Kenya

United Nations University- Institute for Water, Environment and Health (UNU-INWEH) (2024) Global wastewater status. Retrieved from: Global Wastewater Status | United Nations University. Accessed on 5th Nov 2024

UN-Water (2012) UN-Water global annual assessment of sanitation and drinking water (GLAAS) 2012 report: the challenges of extending and sustaining services. Geneva, Switzerland

UN-Water (2022) Progress on wastewater treatment (SDG target 6.3) Retrieved from: https://www.sdg6data.org/en/indicator/6.3.1. Accessed on 18 Nov 2024

UN-Water (2024) Sub-Saharan Africa. Retrieved from: Region | SDG 6 Data. Accessed on 5 Nov 2024

Villarin M, Merel S (2020) Paradigm shifts and current challenges in wastewater management. J Hazard Mater 390:122139. https://doi.org/10.1016/j.jhazmat.2020.122139

Wang H, Wang T, Toure B, Li F (2012a) Protect Lake Victoria through green economy, public participation and good governance. Environ Sci Technol 46(19):10483–10484. https://doi.org/10.1021/es303387v

Wang H, Omosa I, Keller A, Li F (2012b) Ecosystem protection, integrated management and infrastructure are vital for improving water quality in Africa. Environ Sci Technol 46(9):4699–4700. https://doi.org/10.1021/es301430u

Wang H, Wang T, Zhang B, Li F, Toure B, Omosa I, Chiramba T et al (2014) Water and wastewater treatment in Africa—current practices and challenges. Clean 42:1029–1035. https://doi.org/10.1002/clen.201300208

Water Aid (2019) Functionality of wastewater treatment plants in low- and middle-income countries. Water Aid, London, UK

Winpenny J, Heinz I, Koo-Oshima S, Salgot M, Collado J, Hérnandez F et al (2013) Reutilización delAgua en Agricultura: Beneficios para Todos; FAO: Rome, Italy

World Bank Group (2022) Mali: The World Bank is increasing access to water and sanitation services in Bamako. Retrieved from: https://www.worldbank.org/en/news/press-release/2022/11/30/mali-the-world-bank-is-increasing-access-to-water-and-sanitation-services-in-bamako. Accessed on 18 Nov 2024

WWAP (United Nations World Water Assessment Programme) (2017) The United Nations World Water Development Report 2017. Wastewater: the untapped resource. Paris, UNESCO

Yang D, He Y, Wu B, Deng Y, Li M, Yang Q, Liu Y (2020) Drinking water and sanitation conditions are associated with the risk of malaria among children under five years old in sub-Saharan Africa: a logistic regression model analysis of national survey data. J Adv Res 21:1–13. https://doi.org/10.1016/j.jare.2019.09.001

Yesaya M, Tilley E (2021) Sludge bomb: the impending sludge emptying and treatment crisis in Blantyre, Malawi. J Environ Manag 277:111474. https://doi.org/10.1016/j.jenvman.2020.111474

Yongsi H (2009) Wastewater disposal practices: an ecological risk factor for health in young children in Sub-Saharan African cities (case study of Yaoundé in Cameroon). Res J Med Med Sci 4(1):26–41

Chapter 5
Sustainable Management of Wastewater in Sub-Saharan Africa Region

Abstract In this chapter, solutions to manage wastewater (WW) in Sub-Saharan Africa (SSA) region were highlighted. It was evident that improvement of access to water, sanitation and hygiene is key in curtailing mismanagement and discharge of untreated WW to the environment. This can be realized by eradicating open defecation in the region and funding the adoption and use of safe sanitation facilities. The chapter emphasized on the need to adopt green technologies to remediate WW pollutants effectively so that the resource meets predefined local and international standards and does not induce secondary pollution. Improvement of human capacity, tools and equipment to assay for WW pollutants before and after treatment was also recommended as such a measure would improve data collection on prevailing trends in the sector and inform policy and regulatory revisions. Lastly, the incorporation of effluent treatment in a circular economy through resource recovery for energy security and bio-fertilizer production and WW reuse for agricultural irrigation was also suggested. Ultimately, sustainable WW management has positive effects on the environment and public health but its implementation requires strong regulatory, institutional and infrastructural support at all levels of governance.

5.1 Introduction

The management of wastewater (WW) is engraved in the sustainable development goals (SDGs) and in particular, SDG 6.3 whose focus is to half the proportion of untreated WW by increasing reuse and recycling of the resource by 2030 (UN-Habitat 2018). Reuse and recycling efforts reduce WW streams to freshwater systems, protect ecosystems, advance human health and ensure adequate water is available for consumptive uses, which is supportive in realizing other SDGs. However, data paucity on WW treatment is prevalent globally with some sources noting that only 20% of the generated resource is safely discharged into the environment (Lin et al. 2022; Taweesan et al. 2023; UN-Habitat 2023). Without adequate sanitation facilities and protection of freshwater resources from pollution by WW contents, the realization of SDG 6.3 and other related goals will not occur within the stipulated timelines (Boguniewicz and Capodaglio 2017). The prevailing trend is predominant in developing countries such as those of Sub-Saharan Africa (SSA) whose financial, human resources and technological capacity to manage sewage is challenged (Nyika 2022; Omohwovo 2024).

With the established barriers to proper WW management in the developing world despite the projected rise in sewage and effluent generation patterns, there is need to device sustainable solutions to manage the resource and prevent its associated negative effects to the environment, ecosystems and human beings. Ko et al. (2024) noted that WW management globally is intricate due to the many conflicting stakeholders and the complexity of its regulatory structure despite its significance to environmental sustainability and public health. The management of WW is influenced by a number of factors including regulatory provisions but mainly serves to enhance public health and environmental sustainability (Singh et al. 2023).

Lin et al. (2022) highlighted that cost efficient and natural techniques have been established to manage WW based on the intended reuse objective to ensure valorization of the resource in addition to pollutant remediation from effluents to safeguard human health and ecosystems. Sustainable WW treatment efforts require application of advanced technology, a shift from linear to circular management of the resource, legal, policy and institutional modifications to be effective and transformative (Urgina and Milojkovic 2024). Ali and Gujiba (2024) also pointed out the need to continually collect data and monitor the generation and production patterns of WW, their constituents and the health risks that they pose to enable appropriate management measures. This chapter explores the potential solutions to improve WW management in SSA considering the current barriers in management of the resource and using case studies from the region.

5.2 Measures to Manage Wastewater Sustainably

5.2.1 Improvement of Sanitation Facilities

Improved sanitation facilities refers to facilities, which hygienically separate human waste from associated contact (WHO 2024). They include composting toilets, slabbed pit latrines, ventilation-improved pit latrines, pit latrines fitted with septic tanks, pour-flush or flush to piped sewer systems. In SSA region, only 24% of the population had access to safely managed sanitation facilities with 30% and 20% of the total being drawn from urban and rural areas (UN-Water 2024). A further 10% of the total population, 18% and 4% from urban and rural areas respectively had access to basic sanitation services (UN-Water 2024). The rest of the population had access to public-use or shared facilities, pour-flush or flush to elsewhere facilities non-slabbed pit latrines, bucket latrines, open pits, open defecation and hanging latrines, which are not considered as safe (WHO 2024).

5.2 Measures to Manage Wastewater Sustainably

Without adequate sanitation facilities then, generation and discharge of raw WW is the norm. In SSA countries such as Kenya, Lesotho, Uganda, Malawi, Madagascar and Cote d'Ivoire, less than 10% of the population is connected to sewer systems and as such, using unsafe sanitation facilities and services (Andersson et al. 2020). Consequently, WW is polluting land and water resources and efforts to develop sanitation facilities are further complicating WW management approaches. Andersson et al. (2020) noted that modern sanitation facilities that are centralized are incompatible to sustainable development; they are under the control of private connectors and hence difficult to control. Additionally, they are characterized by piping dysfunctions and frequent failures. As such, sanitation and its facility improvement must be prioritized in the developing world just like aspects of education and health care in order to improve WW management particularly, its collection, treatment and reuse. Andersson et al. (2020) noted that sustainable sanitation that addresses piping and technological failures of WW collection will likely contribute to the realization of SDGs 1-15. Dinka and Nyika (2024) had similar views noting that improved access to sanitation and clean water created synergies to realizing other SDGs.

In SSA, countries such as Namibia, Lesotho, Mali, Ethiopia, Zambia, Cameroon, Malawi, South Africa and Nigeria have been reported to have sanitation challenges, which in turn have hindered proper management of WW especially in rural areas (Hlongwa et al. 2024). As such improving such facilities by investing more finances, building a better regulatory framework to manage sanitation failures and prioritizing the sector is advocated for and should be prioritized at national developmental agendas (Hlongwa et al. 2024). Ohwo and Agusomu (2018) highlighted the need to device integrated, inclusive and comprehensive sanitation strategies in SSA that are specific fit for each country to enable healthy living and provision of WW services. In a study evaluating the progress of sanitation service provision in 18 SSA countries, Kanyangarara et al. (2021) noted that sanitation had improved but it was not commensurate with global targets noting that further improvements would preserve public health in the region. Sanitation improvement through decentralized WW treatment and resource recovery in SSA was suggested as a lasting solution to producing effluents that were environmentally acceptable in water-scarce countries of SSA region (Nansubuga et al. 2016).

5.2.2 Innovative and Cost Effective Wastewater Treatment Technologies

The implementation of novel, cost effective, green and innovative water and WW treatment process is key to transforming the SSA journey to realize SDGs as emphasized by Omohwovo (2024). Such efforts require sufficient and sustainable investments in research to innovate local technological solutions to water and WW management that are in line with the predefined international and local standards (Nyiwul 2021). The technological solutions should prioritize on WW disinfection, application of easy and cost effective remediation of effluent pollutants, and the use of natural treatment approaches such as constructed wetlands, combination of bioremediation and pond systems to treat sewage. In addition, energy efficient treatment of WW and prevention of algae growth in centralized treatment plants and in freshwater systems contaminated with WW should be prioritized in the technologies (Wang et al. 2014; Nyika and Dinka 2022). The use of biosorbents from agricultural waste (such as peels and rinds) and biomass to adsorb pollutants from WW through their organic functional groups is also being adopted as a renewable, green, cost-friendly and nature-based technology to treat effluents (Kwikima et al. 2021; Nyika and Dinka 2023).

Although on ad-hoc and small scale extent, the use of environmental friendly, cost effective and innovative WW treatment technologies is being taken up in SSA. In Kenya, constructed wetlands are being used to manage municipal effluents and in the hospitality industry (Makopondo et al. 2020) due to their low energy and maintenance cost. Similar technology has been taken up in Rwanda (Shyaka and Niyonzima 2019) and South Africa (Waly et al. 2022). Chemical disinfection of water using ozonation, chlorination and prexidation and advanced technologies such as ultrafiltration are also being applied in Namibia to manage effluents (Capodaglio 2021). Use of biosorbents such as activated carbon modified biologically or using granules to treat WW for potable use was reported in Namibia (Capadaglio 2021).

In SSA countries such as Uganda, South Africa, Nigeria and Namibia, natural coagulants such as chitosan and *Moringa oleifera* are being used to remediate persistent organic pollutants (POPs) from WW (Ahimbisibwe et al. 2024). The treatment of acid mine drainage WW in South African gold exploration using bioremediation and biostimulation using two bacteria strains proved effective in removal of aluminum and arsenic (Ijoma et al. 2019). Nanotechnology using materials that have surface functional groups to adsorb and absorb WW contaminants is also being developed and tested in management of effluent in SSA (Sambaza et al. 2019; Onu et al. 2023). Plant based biosorbents have been applied to remediate pollutants such as crude oil, mine water, fluorides, antibiotics, pharmaceuticals, heavy metals and dyes from WW of SSA countries such as Botswana, Kenya, South Africa, Zimbabwe, Nigeria, Uganda, Kenya and Tanzania at different removal efficacies (Ngeno et al. 2022). Overall, the application and uptake of the technologies to manage WW needs to be widened to cover large populations and improve their treatment capacities without causing any secondary pollution to the environment (Nyika and Dinka 2023).

5.2.3 *Regulations and Standards*

Many countries in SSA observe the WHO standards for drinking water quality (Omohwovo 2024). However, the efforts to monitor and assess water quality and the effects of WW contamination is hampered by lack of equipment and laboratories in water and WW treatment facilities, poorly equipped laboratories and lack of human resources to conduct the needed tests on WW. Ultimately, water of substandard quality is consumed due to the effect of WW contamination that goes unnoticed. To eradicate such challenges, SSA countries must adhere to international regulations on water quality accompanied by continuous monitoring of water and WW resources. According to Wang et al. (2014), the region and specific countries should establish own-regulations and standards to monitor water and WW quality that are specific to the health and environmental situation and needs in SSA.

Both WHO and European Union (EU) have developed standards on WW treatment and reuse based on the levels of nutrient and organic parameters before reuse of the resource, especially for agriculture (Onu et al. 2023). The acceptable standards and regulations depend on the source of WW, the approach of discharge and the reuse objective. Specific SSA countries such as Rwanda, Nigeria and Kenya have specific standards for monitoring WW before its safe discharge and reuse. In Kenya and Nigeria, the regulations and standards are highlighted in the environment management coordination act (EMCA) and the federal environmental protection agency (FEPA) act, respectively. However, specific parameters in treated WW in the countries in addition to Rwandan WW discharge standards was reported to be higher than the national standards and also the WHO standards (Aniyikaiye et al. 2019; Durotoye et al. 2018; Egwuonwu et al. 2012; Osin et al. 2017; Theoneste et al. 2020). Such trends are common in SSA countries and are indicative of ineffective effluent treatment and as such, the need for continuous monitoring of WW against the predefined standards is a priority (Onu et al. 2023).

In Uganda, standards on WW treatment and discharge exist but complying with them is hindered by the use of conventional management approaches as noted by Bateganya et al. (2015). Therefore, it is imperative for SSA region to raise its health standards to enable continuous water and WW monitoring, compliance with existence local and international regulations and standards on treatment and discharge in addition to developing a strong WW safety plan that prioritizes human health and environmental sustainability (Onu et al. 2021). Authorities in the WW sector should put concerted efforts to address the deficits and failures of the treatment technologies applied and their associated operations and maintenance (Bateganya et al. 2015).

5.2.4 Enhancement of Human Capacity

In SSA, the human capacity to manage WW is challenged by a number of factors. The regions suffers from a deficiency of technicians, operators and engineers with sufficient expertise to construct, design, make operational and maintain WW infrastructure and treatment facilities. The trend was observed in parts of Durban (South Africa) and Windhoek (Namibia), whereby sewage management was problematic (Olbrisch 2006; Tseole et al. 2022). As such, the staff involved in cycling WW should receive adequate training to enhance their skills and knowledge in managing the resource in addition to equipping them with advanced technology to operate and maintain functional treatment facilities (Omohwovo 2024). Advances to improve human capacity should be coupled with enhanced technical knowledge to the staff, elimination of outdated WW treatment technologies that perform sub-optimally and result to extensive environmental degradation as well as financial support from both the government and the private sector (Tseole et al. 2022).

The region also suffers from institutional incapacity to manage WW characterized by weak institutional frameworks and lack of resources to undertake transformational actions in the management of the resource. With weak institutions, implementation of WW regulations is uncoordinated and poor, lacks planning and inadequate involvement of communities in decision-making as established in Uganda (Kulanyi et al. 2021). Ultimately, the management of WW and the ability to provide water, hygiene and sanitation services are negatively affected. Libya for example has sewage treatment plants adequate to manage municipal WW from Tobruk but lack of institutional framework for the undertakings results to process inefficiency (Fanack Water 2020).

Present institutions also do not enforce existent WW discharge regulations adequately (Fanack Water 2020). The existent of pre-independence policies in modern day and lack of their revision further impede provision of WW services among the minority and marginalized groups as is the case in South Africa (Maphumulo and Bhengu 2019). Therefore, the need to build strong institutions supported by effective regulatory and policy frameworks to incentivize and support WW treatment, reuse and recovery among other sustainable practices should be prioritized. In SSA countries of Zambia, Kenya and Tanzania, stronger institutional frameworks have been formulated to manage WW but their effectiveness must be promoted through better regulatory arrangements, improved preparedness, monitoring and better capacity building (UN-Habitat 2023). In Dhaka, Senegal, the management of WW onsite has improved significantly following a strengthened institutional capacity to manage the resource (UN-Habitat 2023).

5.2.5 Governance Improvement in the Wastewater Sector

Governance in the WW sector of SSA suffers from several shortcomings. The responsibility to manage the resource is spread across several ministries whose coordination is poos and the roles of each ministry are not clearly defined (Omhwovo 2024). Similarly, the providers of WW services are multiple and drawn from both the public and private sectors, which makes resource management unsustainable (Medland et al. 2016; Sanusi et al. 2023). Consequently, there is lack of cooperation from the involved parties and a duplication of roles in the management of WW leading to inefficiency and ineffectiveness. The trend was reported in the Democratic Republic of Congo where coordination of ministries with different and sometimes, duplicated roles to manage WW was attributable to ineffectiveness, the increase in malnutrition and a high prevalence of waterborne diseases (Omohwovo 2024). Similar tendencies have been reported in SSA countries of Uganda, Cameroon, Ghana, Senegal, Mozambique, Rwanda and Kenya (Medland et al. 2016).

5.2 Measures to Manage Wastewater Sustainably

Moving forward, it is imperative for the private and public sector to have coordinated measures of managing WW that promote sustainability, efficiency and equity by applying adaptive strategies and effective governance (Sanusi et al. 2023). Ekane et al. (2014) recommended the need to revise institutional and governance frameworks through memorandum of understanding among ministries in development and implementation of WW management plans to enable coherence and reverse the fragmentation effects. Omohwovo (2024) noted that fragmentation of WW management roles can be eradicated using effective structures that prioritize on accountability, application of strong policy frameworks and promote capacity building in the entire water sector for sustainable development.

The political will to improve the WW sector in SSA is often lacking (Dos Santos et al. 2017). This is because WW management is not considered a vote winning agenda and hence, is given low priority. In addition, the public who mainly generate and interact with WW are not involved in decisions on the management of the resource (Wang et al. 2014). As such, WW management awareness is limited at community level. The low priority and lack of public engagement in WW management is further exacerbated by rapid expansion of urban areas that makes it complicated to enhance coverage of services even with political lobbying (Dos Santos et al. 2017). To improve WW management, public participation is essential since people are the ones who drive agendas that are of priority to them including WW management. In addition, providing education and training on WW management using innovative and local practices, its importance and consequences of mismanagement is advocated to improve awareness at community level in SSA as well as promote the understanding of the water, health and energy nexus in reference to scientific WW management (Wang et al. 2014; Omohwovo 2024). In rural Uganda, a combination of community awareness programs and public participation in local planning for WW management reformed the local politicians to get involved in the resource's management and sanitation improvement proactively (Quin et al. 2011).

5.2.6 *Improve Wastewater Monitoring, Data Collection and Sharing*

Data on WW production, collection, treatment and reuse in SSA is hardly collected. Such tendencies are associated with limited infrastructure that makes it impossible to collect and assay representative samples. The cost of designing, operating and maintaining WW monitoring systems is high in the countries, which are already strained economically. The countries also lack technical expertise to collect data on WW trends, analyze it and interpret the information for policy revision. Additionally, the region does not have formal standards on data sharing and protection on such trends. Owing to the challenges, WW data is scattered, unavailable, unused if available and unreported in most cases, which has challenged resource management (Qadir et al. 2020; Jones et al. 2021; Ali and Gujiba 2024; Drechsel et al. 2024).

To improve the availability and use of WW data, there is a need to improve existing infrastructure including sewer connections for all and collection of more WW to treatment plants. Such infrastructural improvements include engaging local communities in data collection, setting up mobile laboratories for WW sample collection and assays, collaborating with public, private, local and international organizations involved in research on WW, its effects and better management approaches (Manirambona et al. 2024). To promote data sharing, digital platforms can be used and standardized protocols of sample collection, preparation, assay and reporting need to be established to improve WW data comparability, reproducibility and quality. Such efforts have been established in several SSA countries including Mozambique, South Africa, Kenya, Nigeria and Uganda especially post-COVID-19 pandemic (Manirambona et al. 2024). In Mozambique, equipping laboratories with WW monitoring equipment and reagents as well as educating laboratory staff bettered the capacity to collect WW data (WHO Africa 2008). In countries such as Nigeria and South Africa, WW-based surveillance of has resulted to better management of poliovirus and norovirus, respectively (Manirambona et al. 2024). With quality WW data, waterborne diseases surveillance can be improved and the environmental monitoring can be bettered for affirmative action.

5.2.7 Enhanced Resource Reuse and Recovery from Wastewater

The prevailing trends of population growth and expansion of urban centers in SSA region have promoted water use and subsequently, WW generation and its associated pollution effect (Wang et al. 2014). The management of WW in the region largely influences economic trends and has the propensity to induce negative and positive effects (Smol et al. 2020). Scientific management of the resource can induce advantages of environmental sustainability by promoting water security. For this reason, modern research agendas are advocating for the infusion of WW management in a circular economy (CE). In such CE advances, WW will no longer be considered as waste that requires elimination but rather a raw material to make new and valuable products (Neczaj and Grosser 2018). In a CE, WW can be processed for energy generation in the form of biofuels such as biogas or for bio-fertilizer production through the recovery of nutrients such as nitrogen and phosphorous (Neczaj and Grosser 2018; Hossain et al. 2020). Additionally the resource can be treated and reused for various consumptive uses without causing any negative environmental effects (Ghasemi-Zaniani et al. 2017). Valuable components of WW such as plastics and metals are recoverable from the resource for recycling as has been done in developed countries such as the United States (Mulchandani and Westerhoff 2016).

Initiatives geared towards incorporating WW in the CE are present in SSA countries and have been transformative in sustainable management of the resource. Sanergy Limited, which is a company in Kenya, is known for collecting black water

5.2 Measures to Manage Wastewater Sustainably

and fecal sludge from informal settlements and using it to manufacture organic fertilizers (Good Family 2024). Consequently, the company is contributing positively to improving sanitation, hygiene and access to clean water in the slums through reduced environmental pollution. The Dakar WW treatment facility of Senegal has been a model example on how WW can be managed in a CE (World Global Practice 2021). In the plant, WW treatment has been enhanced and in the processes, fertilizers have been recovered, methane gas has been recovered as a source of energy, effluents have been reused for agriculture and ultimately, the ecosystem within the facility's vicinity has been restored.

In SSA countries of Nigeria, Ghana, Uganda, South Africa, Gabon, Zimbabwe and Cameroon organic waste and wastewater was converted to bio-fertilizer and biogas (Rubagumya et al. 2023). The potential for such CE practices ranged from 11.1 to 306.26 million tons a year in the involved countries with Ghana having the greatest production potential. Bio-latrines that process Blackwater, fecal and organic waste through bio-digestion have been developed and built as green measures to improve sanitation and process WW in Zambian schools courtesy of the national water and sanitation association (Relief Web 2016). To integrate WW effectively in a CE, there is need for SSA countries to invest in their infrastructure, adopt innovative technologies, engage communities, public and private partners in decisions on WW management, build capacity to manage the resource and culture strong institutional, regulatory and policy frameworks to support the WW cycle and its evaluation and monitoring (AUDA 2024).

5.2.8 Reliable Energy Supply

Unreliable energy supply characterized by blackouts significantly hinder WW management in many African countries (Wang et al. 2014; Omohwovo 2024). Without a constant power supply, WW treatment and efforts to recover its valuable components are insufficient as has been reported in Nigeria (Oloruntoba and Alabi 2019) and SSA in general (Wang et al. 2012). The power supply problem is worsened by the use of the traditional power grid supply and the application of high-energy demanding treatment technologies such as aeration. Additionally, the countries have endemic financial problems that act as barriers to investing in green-energy to treat WW and improve its associated infrastructure and technologies (Rugaimukamu et al. 2022).

Moving forward, the region must invest in more energy and land efficient WW treatment approaches such as nanofiltration in place of conventional energy intensive technologies. Such a move will save on the use of fossil fuels and control the apparent greenhouse gas emissions associated with such energy uses (Rugaimukamu et al. 2022). The use of green energy including solar photovoltaics and wind turbines in powering the treatment of WW is advocated for considering the great potential the energy sources have yet they are largely unexploited in the region (Wang et al. 2014; Rugaimukamu et al. 2022).

Valorization of WW treatment approaches particularly anaerobic digestion can also be done to produce energy in the form of biogas, which in turn can generate heat and electricity for not only further treatment but also other uses (Wang et al. 2014). Such advances have been taken up in Australia and were found to be imperative in recovery of WW components for enhance energy security (Rugaimukamu et al. 2022). In South Africa, Outeniqua, Kraanfontein, Hartenbos and Gwaiing WW treatment plants rely on solar photovoltaics for power supply, which has reduced overreliance on national grid electricity significantly (George Municipality 2023; Chandak 2024). In areas of Nyeri County and Gorge Farm in Kenya, fecal sludge and agro-wastes have been used in generating biogas, which is useful in providing heat and electricity (Lumadede et al. 2021). By adopting green energy sources, WW treatment can be sustainable and the resultant effluents can be used as valuable resources to enhance power security in SSA.

5.3 Conclusion

This chapter highlighted the importance of taking corrective measures in the SSA WW sector to manage it and its associated environmental and public health effects better. Key recommendations made include improving sanitation facilities by avoiding open defecation and unsafe sanitation approaches that lead to more WW generation and its discharge in freshwater systems, which results to extensive environmental pollution. The chapter also suggested the need to advance to greener and advanced treatment technologies rather than traditional ones that treat WW sub-optimally without adhering to the predefined local and international standards. Improved governance in the WW sector that includes political willingness to fund and prioritize WW infrastructural improvement, involve the public in participatory management of the resource and create awareness and positive attitudes on the management and use of recycled effluents was suggested. Such a move will improve resource recovery from WW, water scarcity resilience, energy security and enhance agricultural potential of SSA using treated effluent. The use of reliable and green energy in treatment of sewage will also enhance capacity and further promote sustainable environments. To implement these measures effectively, strong institutions and support from the private, public, local and international sectors are prerequisites. In conclusion, the overhaul of the WW sector in SSA region is a need of the hour priority towards realizing sustainable development.

References

African Union Development Agency, AUDA (2024) Creating an enabling environment and utilizing technologies for sewage and wastewater management in Africa. https://www.nepad.org/blog/creating-enabling-environment-and-utilizing-technologies-sewage-and-wastewater-management. Accessed 5th Dec 2024

Ahimbisibwe M, Ssebugere P, Nagawa C, Matovu H, Nakawuka P, Isa K et al (2024) Potential of natural coagulants for bioremediation of persistent organic pollutants in wastewater in Sub-Saharan Africa: a review. Arch Agri Environ Sci 9(2):385–404. https://doi.org/10.26832/24566632.2024.0902026

Andersson K, Rosemarin A, Lamizana B, Kvarnström E, McConville J, Seidu R, Dickin S, Trimmer C (2020) Sanitation, wastewater management and sustainability: from waste disposal to resource recovery, 2nd edn. United Nations Environment Programme and Stockholm Environment Institute, Nairobi and Stockholm

Aniyikaiye T, Oluseyi T, Odiyo J, Edokpayi J (2019) Physico-chemical analysis of wastewater discharge from selected pain industries in Lagos, Nigeria. Int J Environ Res Public Health 16(7):1235. https://doi.org/10.3390/ijerph16071235

Ali A, Gujiba U (2024) Household wastewater management in sub-Saharan Africa: a review. Discov Water 4:6. https://doi.org/10.1007/s43832-024-00060-6

Bateganya N, Nakalanzi D, Babu M, Hein T (2015) Buffering municipal wastewater pollution using urban wetlands in sub-Saharan Africa: a case of Masaka municipality, Uganda. Environ Technol 36(17):2149–2160. https://doi.org/10.1080/09593330.2015.1023363

Boguniewicz J, Capodaglio A (2017) Sustainable wastewater treatment solutions for rural communities: public (centralized) or individual (on-site). Case study. Econ Environ Stud 17(4):1103–1119. https://doi.org/10.25167/ees.2017.44.29

Capodaglio A (2021) Fit-for-purpose urban wastewater reuse: analysis of issues and available technologies for sustainable multiple barrier approaches. Crit Rev Environ Sci Technol 51(15):1619–1666. https://doi.org/10.1080/10643389.2020.1763231

Chandak P (2024) Cape Town completes solar project at Kraaifontein wastewater plant. Solar Quarter, Mumbai, India

Dinka M, Nyika J (2024) SDG 6 progress analyses in sub-Saharan Africa from 2015–2020: the need for urgent action. Discov Water 4:39. https://doi.org/10.1007/s43832-024-00099-5

Dos Santos S, Adams E, Neville G, Wada Y, Sherbinin A, Bernhardt M, Adamo S (2017) Urban growth and water access in Sub-Saharan Africa: progress, challenges and emerging research directions. Sci Total Environ 607–608:497–508. https://doi.org/10.1016/j.scitotenv.2017.06.157

Drechsel P, Bartram J, Qadir M, Medlicott K (2024) The challenge of supporting and monitoring safe wastewater use in agriculture in LMIC. NPJ Clean Water 6(7):1–3. https://doi.org/10.1038/s41545-024-00364-z

Durotoye T, Adeyemi A, Omole D, Onakunle O (2018) Impact assessment of wastewater discharge from a textile industry in Lagos, Nigeria. Cogent Eng 5(1):1531687. https://doi.org/10.1080/23311916.2018.1531687

Egwuonwu C, Uzoije A, Okafor V, Ezeanya N, Nwachukwu M (2012) Evaluation of the effects of industrial wastewater discharge on surface water: a case study of Nigeria Breweries PLC, Enugu. Green J Phys Sci 2(3):56–63

Ekane N, Nykvist B, Kjellen M, Noel S, Weitz N (2014) Multi-level sanitation governance; understanding and overcoming sanitation sector in Sub-Saharan Africa. Waterlines 33(3):242–256. https://doi.org/10.3362/1756-3488.2014.024

Fanack Water (2020) Water management and challenges in Libya. https://water.fanack.com/libya/water-management-in-libya/. Accessed 3 Dec 2024

George Municipality (2023) George unveils the Outeniqua wastewater treatment works for solar energy plant. https://www.george.gov.za/george-unveils-the-outeniqua-waster-water-treatment-works-solar-energy-plant/. Accessed 4 Dec 2024

Ghasemi-Zaniani M, Eslamian S, Ostad-Ali Askari K, Singh V (2017) Irrigation with wastewater treated by constructed wetlands. Int J Res Stud Agric Sci 3(11):18–34. https://doi.org/10.20431/2454-6224.0311002

Good Family (2024) Sanergy-pioneering a zero organic waste economy. https://good-search.org/about/en/project/sanergy/. Accessed 4 Dec 2024

Hlongwa N, Nkomo S, Desai S (2024) Barriers to water, sanitation and hygiene in Sub-Saharan Africa: a mini review. J Water Sanit Hyg Dev 14(7):497. https://doi.org/10.2166/washdev.2024.266

Hossain N, Bhuiyan M, Pramanik B, Nizamuddin S, Griffin G (2020) Waste materials for wastewater treatment and waste adsorbents for biofuel and cement supplement applications: a critical review. J Clean Prod 255:120261. https://doi.org/10.1016/j.jclepro.2020.120261

Ijoma G, Selvarajan R, Oyourou J, Sibanda T, Matambo T, Monanga A et al (2019) Exploring the application of biostimulation strategy for bacteria in the bioremediation of industrial effluent. Ann Microbiol 69:541–551. https://doi.org/10.1007/s13213-019-1443-6

Jones E, Van Vliet M, Qadir M, Bierkens M (2021) Country-level and gridded estimates of wastewater production, collection, treatment and reuse. Earth Syst Sci Data 13(2):237–254. https://doi.org/10.5194/essd-13-237-2021

Kanyangarara M, Allen S, Jiwani S, Fuente D (2021) Access to water, sanitation and hygiene services in health facilities in sub-Saharan Africa 2013–2018: results of health facility surveys and implications for COVID-19 transmission. BMC Health Serv Res 21:601. https://doi.org/10.1186/s12913-021-06515-z

Ko D, Norton J, Daigger G (2024) Wastewater management decision-making: a literature review and synthesis. Water Environ Res 96(4):e11024. https://doi.org/10.1002/wer.11024

Kulanyi R, Jonga M, Vos E, Van Soelen S (2021) Barriers to inclusion in the WASH sector: insights from Uganda. Int J Biol Chem Sci 15(7):76–81. https://doi.org/10.4314/ijbcs.v15i7.8S

Kwikima M, Mateso S, Chebude Y (2021) Potentials of agricultural wastes as the ultimate alternative adsorbent for cadmium removal from wastewater. A review. Sci Afr 13:e00934. https://doi.org/10.1016/j.sciaf.2021.e00934

Lin L, Yang H, Xu X (2022) Effects of water pollution on human health and disease heterogeneity: a review. Front Environ Sci 10:880246. https://doi.org/10.3389/fenvs.2022.880246

Lumadede M, Wangai L, Kwach S, Khalifa J, Mbithi V (2021) Biogas technology in Kenya: a review. J Environ Sci Comp Sci Eng Technol 10(3):369–381. https://doi.org/10.24214/jecet.A.10.3.36981

Makopondo R, Rotich L, Kamau C (2020) Potential use and challenges of constructed wetlands for wastewater treatment and conservation in game lodges and resorts in Kenya. Sci World J 1–9. https://doi.org/10.1155/2020/9184192

Manirambona E, Lucero D, Shomuyiwa D, Denkyira A, Okesanya O, Haruna U et al (2024) Harnessing wastewater-based surveillance (WBS) in Africa: a historic turning point towards strengthening the pandemic control. Discov Water 4:9. https://doi.org/10.1007/s43832-024-00066-0

Maphumulo W, Bhengu B (2019) Challenges of quality improvement in the healthcare of South Africa post-apartheid: a critical review. Curations 42(1):1–9. https://doi.org/10.4102/curationis.v42i1.1901

Medland L, Scott R, Cotton A (2016) Achieving sustainable sanitation chains though better informed and more systematic improvements: lessons from multi-city research in Sub-Saharan Africa. Environ Sci Water Res Technol 2:492–501. https://doi.org/10.1039/c5ew00255a

Mulchandani A, Westerhoff P (2016) Recovery opportunities for metals and energy from sewage sludges. Bioresour Technol 215:215–226. https://doi.org/10.1016/j.biortech.2016.03.075

Nansubuga I, Banadda N, Verstraete W, Rabaey K (2016) A review of sustainable sanitation systems in Africa. Rev Environ Sci Biotechnol 15:465–478. https://doi.org/10.1007/s11157-016-9400-3

Neczaj E, Grosser A (2018) Circular economy in wastewater treatment plant–challenges and barriers. Paper presented at the Multidisciplinary Digital Publishing Institute Proceedings

References

Ngeno E, Mbuci K, Necibi M, Shikuku V, Olisah C, Ongulu R et al (2022) Sustainable re-utilization of waste materials as adsorbents for water and wastewater treatment in Africa: recent studies, research gaps, and way forward for emerging economies. Environ Adv 9:100282. https://doi.org/10.1016/j.envadv.2022.100282

Nyika J (2022) Wastewater for agricultural production, benefits, risks, and limitations. In: Chatoui H, Merzouki M, Moummou H, Tilaoui M, Saadaoui N, Brhich A (eds) Nutrition and human health. Springer, Cham. https://doi.org/10.1007/978-3-030-93971-7_6

Nyika J, Dinka M (2022) A mini-review on wastewater treatment through bioremediation towards enhanced field applications of the technology. AIMS Environ Sci 9(4):403–431. https://doi.org/10.3934/environsci.2022025

Nyika J, Dinka M (2023) The application of microorganism-derived biosorbents in the removal of heavy metals and dyes. In: Ukhurebor K, Aigbe U, Onyancha R (eds) Adsorption applications for environmental sustainability. IOP Publishing Ltd.

Nyiwul L (2021) Innovation and adaptation to climate change: evidence from the water sector in Africa. J Clean Prod 298:126859. https://doi.org/10.1016/j.jclepro.2021.126859

Ohwo O, Agusomu T (2018) Assessment of water, sanitation and hygiene services in Sub-Saharan Africa. Eur Sci J 14(35):308–326. https://doi.org/10.19044/esj.2018.v14n35p308

Olbrisch S (2006) Optimizing sewage management in urban settlements in Sub-Saharan Africa: a contribution to urban development planning. Sustainable Sanitation Alliance, German

Oloruntoba S, Alabi T (2019) Domestic wastewater reclamation and Reuse in Nigeria: a case study of some selected treatment Abuja and lagos. J Future Eng Technol 15(1):1–10. https://doi.org/10.26634/jfet.15.1.14953

Onu M, Ayeleru O, Modekwe H, Oboirien B, Olubambi P (2021) Water and wastewater safety plan in Sub-Saharan Africa. In: Dehghani M, Tyagi I, Karri R, Scholz M (eds) Water, the environment and the sustainable development goals. Elsevier. https://doi.org/10.1016/B978-0-443-15354-9.00019-0

Omohwovo E (2024) Wastewater management in Africa: challenges and recommendations. Environ Health Insights 18:1–6. https://doi.org/10.1177/11786302241289681

Onu M, Ayeleru O, Oboirien B, Olubambi P (2023) Challenges of wastewater generation and management in sub-Saharan Africa: a review. Environ Chall 11:100686. https://doi.org/10.1016/j.envc.2023.100686

Osin O, Yu T, Lin S (2017) Oil refinery wastewater treatment in the Niger Delta, Nigeria: current practices, challenges, and recommendations. Environ Sci Pollut Res 24(28):22730–22740. https://doi.org/10.1007/s11356-017-0009-z

Qadir M, Drechsel P, Jiménez B, Kim Y, Pramanik A, Mehta P et al (2020) Global and regional potential of wastewater as a water, nutrient and energy source. Nat Resour Forum 44:40–51. https://doi.org/10.1111/1477-8947.12187

Quin A, Balfors B, Kjellen M (2011) How to "walk the talk": the perspectives of sector staff on implementation of the rural water supply program in Uganda. Nat Resour Forum 35(4):269–282. https://doi.org/10.1111/j.1477-8947.2011.01401.x

Relief Web (2016) Inspiring innovation to deliver clean water and sanitation in Zambia. https://reliefweb.int/report/zambia/inspiring-innovation-deliver-clean-water-and-sanitation-zambia. Accessed 5 Dec 2024

Rubagumya I, Komakech A, Kabenge I, Kiggundu N (2023) Potential of organic waste to energy and bio-fertilizer production in Sub-Saharan Africa: a review. Waste Dispos Sustain Energy 5:259–267. https://doi.org/10.1007/s42768-022-00131-1

Rugaimukamu Q, Wang H, Huang R, Xie L (2022) Wastewater treatment needs more attention. Nat Afr. https://doi.org/10.1038/d44148-022-00071-2

Sambaza S, Maity A, Pillay K (2019) Enhanced degradation of BPA in water by PANI supported Ag/TiO2 nanocomposite under UV and visible light. J Environ Chem Eng 7(1):102880. https://doi.org/10.1016/j.jece.2019.102880

Sanusi O, Oke M, Bello M (2023) Water entrepreneurship and financialisation: complexities for the attainment of SDG in Sub-Saharan Africa. Heliyon 9(11):e20859. https://doi.org/10.1016/j.heliyon.2023.e20859

Shyaka E, Niyonzima E (2019) Sustainable design of wastewater treatment plant in Kibiligi Quater. Int J Appl Eng Res 14(22):4105–4111

Singh B, Chakraborty A, Sehgal R (2023) A systematic review of industrial wastewater management: evaluating challenges and enablers. J Environ Manage 348:119230. https://doi.org/10.1016/j.jenvman.2023.119230

Smol M, Adam C, Preisner M (2020) Circular economy model framework in the European water and wastewater sector. J Mater Cycles Waste Manage 22(3):682–697. https://doi.org/10.1007/s10163-019-00960-z

Taweesan A, Kanabkaew T, Surinkul N, Polprasert C (2023) Convenient solutions to inconvenient truth: domestic wastewater management-based approaches to sustainable development goal no.6. Environ Sustain Indic 18:100255. https://doi.org/10.1016/j.indic.2023.100255

Theoneste S, Vincent N, Xavier N (2020) The effluent quality discharged and its impacts on the receiving environment case of Kacyiru sewerage treatment plant, Kigali, Rwanda. Int J Environ Agric Res 6(11):20–29

Tseole N, Mindu T, Kalinda C, Chimbari M (2022) Barriers and facilitators to water, sanitation and hygiene (WaSH) practices in Southern Africa: a scoping review. PLoS ONE 17(8):e0271726. https://doi.org/10.1371/journal.pone.0271726

UN-Habitat (2018) SDG indicator 6.3.1 training module: safe wastewater treatment. United Nations Human Settlement Programme (UN-Habitat), Nairobi, Kenya

UN-Habitat (2023) Global report on sanitation and wastewater management in cities and human settlements. Nairobi, Kenya

UN-Water (2024) Sub-Saharan Africa. Retrieved from: Region | SDG 6 Data. Accessed 27 Nov 2024

Urgina M, Milojkovic J (2024) Advances in wastewater treatment, 2024. Energies 17(6):1400. https://doi.org/10.3390/en17061400

Waly M, Ahmed T, Abunada Z, Mickovski S, Thomson C (2022) Constructed wetland for sustainable and low-cost wastewater treatment. Land 11(9):1388. https://doi.org/10.3390/land11091388

Wang H, Omosa I, Keller A, Li F (2012) Ecosystem protection, integrated management and infrastructure are vital for improving water quality in Africa. Environ Sci Technol 46(9):4699–4700. https://doi.org/10.1021/es301430u

Wang H, Wang T, Zhang B, Li F, Toure B, Omosa I, Chiramba T et al (2014) Water and wastewater treatment in Africa–current practices and challenges. Clean 42:1029–1035. https://doi.org/10.1002/clen.201300208

Water Global Practice (2021) Water in circular economy and resilience (WICER); the case of Dakar, Senegal. Global Water Security and Sanitation Partnership and World Bank Group

WHO (2024) Improved sanitation facilities and drinking water sources. https://www.who.int/data/nutrition/nlis/info/improved-sanitation-facilities-and-drinking-water-sources. Accessed 27 Nov 2024

WHO Africa (2008) The Maputo declaration on strengthening of laboratory systems. https://www.who.int/publications/m/item/the-maputo-declaration-on-strengthening-of-laboratory-systems. Accessed 4 Dec 2024

The manufacturer's authorised representative in the EU is Springer Nature Customer Service Centre GmbH, Europaplatz 3, 69115 Heidelberg, Germany. If you have any concerns regarding our products, please contact ProductSafety@springernature.com

Printed and bound by CPI Group (UK) Ltd, Croydon, CR0 4YY
26/03/2026
02078971-0003